the
greengrocer

the greengrocer

LEANNE KITCHEN

MURDOCH BOOKS

contents

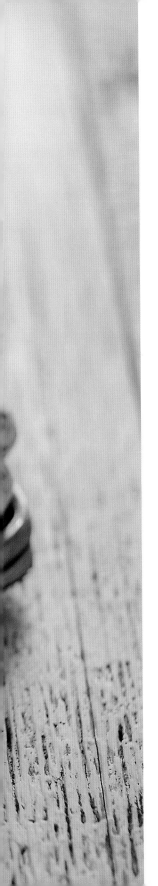

introduction

The fact our society is organised into villages, towns and cities owes much, if not everything, to agriculture. Millennia ago, humans were hunters, gatherers and fishers, living in small, nomadic groups ranging in search of food. Some 10,000 years ago, agriculture sprang up in several regions of the world, providing valuable staple crops such as wheat, peas, barley, lentils and oats.

When cow, horse and donkey power were harnessed, agriculture became a little less dependent upon human labour. The 1700s brought all the mechanical advances of the Industrial Revolution, and also saw the introduction to Europe of high-yielding New World crops such as corn and potatoes, which had an enormous impact on farming and on food supply. Throughout the following centuries, improvements to farming hardware, labour-saving machinery and cultivation methods greatly increased crop yields and food supplies — and world populations have grown accordingly.

The advent of refrigeration, highly sophisticated transportation networks, scientific know-how and production on an industrial scale have made modern agriculture a global business. Go to the supermarket today and you might find asparagus from Kenya, oranges from Israel or lychees from China. A wide range of fruits, vegetables and herbs are now available all year round, sold so far from the farm gate that it's almost hard to believe they came out of the soil at all.

THE RHYTHM OF THE SEASONS

In the past, when our very existence was bound up in the land and what it produced for us to eat, bonds between humans and the earth ran strong. It was understood that the weather, the unpredictable forces of nature and the seasons of the year should be respected and even venerated.

How things have changed. Our fruits and vegetables have been hybridised to better resist disease, ripen early, or grow to a standard size and colour — yet despite the undoubted convenience this offers, the trade-off is often a loss of flavour and character.

Today there is a growing push to 'eat locally and eat seasonally'. More of us are avoiding non-seasonal produce in favour of the bounty each season offers — eating fresh tomatoes and strawberries only in summer, and enjoying asparagus and peas in spring, for example. Celery still tastes at its crispy best when harvested in winter, while basil is at its height of peppery sweetness in summer. Some produce, however, doesn't appreciably decline in quality when grown (often in greenhouses) year round. Others, like potatoes and pumpkin (winter squash), naturally store well and are not harmed by long keeping.

As well as the environmental and health benefits of buying locally grown in-season produce, there are obvious benefits to the household budget — seasonal vegetables and fruits are not only at their most flavoursome, but in their plentitude are also at their cheapest.

Due to health concerns and environmental considerations, more consumers are seeking out organically grown produce, raised without the assistance of chemical pesticides and fertilisers. Foods that are grown organically also tend to taste better, say many.

In its purest form, organic farming is a holistic approach to agriculture that has at its heart the health of the soil and the ongoing sustainability of land use. Government organisations in many countries issue special certification to producers only if they comply with the rigorous standards required to be labelled 'organic'. Such certification is usually displayed on the product at its point of sale so consumers can be sure they are buying the 'real thing'. Ironically, before mass-produced nitrate fertilisers became available around World War I, pretty much all fruits and vegetables were 'organic'.

In the 1980s, the ability to alter food plants at a genetic level revolutionised agricultural production. Genetic engineering, or genetic modification (GM) as it is also called, enables scientists to 'tweak' genes to achieve various 'improvements' such as increasing the plant's yield, disease resistance or nutritional profile, or extending its shelf life. There are several ways this can be done. One involves isolating the DNA from one organism or plant, modifying it in a laboratory, then inserting a new gene or genes into the recipient plant. Other methods involve increasing or decreasing the amount of a gene already present in a plant, removing it altogether or changing its position in the genetic structure.

Numerous scientific studies have been conducted into the safety and ethics of genetic modification, and many scientists say GM foods are no more 'tampered' with than any other modern fruits and vegetables produced by the centuries-old practice of selective breeding. Supporters make many good claims for the positive impact GM foods can have — for instance, plants can grow in harsher conditions than they naturally would, opening up possibilities for agriculture in impoverished or hungry regions of the world, while strains of rice developed to contain high levels of vitamin A would have indisputable benefits in undernourished populations.

Detractors call GM foods 'Frankenfoods' and are opposed to them on ethical, social or scientific grounds. They say there is no way of knowing the long-term effects of consuming GM foods, and are concerned that GM plant stocks could potentially contaminate wild and conventional plant stocks and forever change their DNA.

Most countries that produce and sell GM foods are obliged to label those foods as such; the crops currently most subject to genetic modification include corn, rice, wheat, soya beans and canola. The United States grows and sells more GM foods than any other country. With the practice still in its relative infancy, it is hard to predict how widespread and accepted it will become. Nevertheless, with products labelled as such, consumers will be able to exercise the power of choice.

KITCHEN BASICS

Fruit and vegetable cookery spans the entire menu. The equipment you have on hand for your general cooking will suffice for most recipes involving fresh produce, although having a few specialised items can make some of the more laborious or tricky jobs a cinch.

CHARGRILL PAN

The beauty of a chargrill pan is that it enables you to 'barbecue' indoors. Essentially the 'plate' of the pan is like a griddle, but with ridges over the base to keep the food above the heat source. These ridges sear the food and mark it as a barbecue would, but the flavour is not quite the same as that on a true barbecue as there is no contact with flame or smoke. These pans are excellent for vegetables that might otherwise be incinerated on a barbecue, such as zucchini (courgette), eggplant (aubergine) and capsicum (pepper).

Some chargrill pans are almost flat; others have raised sides, like a frying pan. Some are round, some square and others rectangular. Choose one with a heavy metal construction; cast iron is excellent. You need a reasonably long handle so you don't have to get too close to the heat; some don't have handles at all and require the use of tongs and oven mitts to be removed from the heat. A big consideration should be the capacity of the pan — be guided by how much food you'll want to cook at a time. One that is too small can see you hovering over a hot stove for much longer than necessary.

CHERRY (OR OLIVE) PITTER

There's really no easy way to remove the stones from cherries — other than by using one of these specialty items. (A cherry pitter is very effective at removing the stones from olives too!) You just place each cherry in the little round holder, stem end facing you, then press down on the spring-loaded handle. This pushes a fine plunger through the cherry, expelling the stone through a hole at the base of the holder. Easy! Choose a good, metal pitter that feels comfortable in your hand and is effortless to use.

CHOPPING BOARD

So much in the kitchen starts here, at the chopping board. When preparing fruits and vegetables, you'll need a very good chopping board, if not several — it is handy to have a small board solely for preparing garlic, as its pungency is hard to remove and can easily taint delicately flavoured foods.

Plastic, rubber or wood are all ideal materials as they are soft enough to absorb the impact of a knife and not blunt its blade, yet solid enough to be stable on a work surface. Some cooks favour wood as it is a natural substance; there is debate though over whether wood can harbour bacteria. High-density thermoplastic boards are much used by professional cooks — unlike wooden boards they won't chip, crack, swell or warp, won't absorb moisture or taint from food and are easily cleaned by bleaching. Professional cooks also favour boards with hard rubber surfaces as they are non-porous, stable on the bench, easier on knife edges and easier to clean than wooden boards.

Choose boards that are durable, thick, stable, and probably much bigger than you think you'll need. Have at least one very large board for recipes that require you to chop large quantities of fresh produce.

COLANDER

A colander is so useful for washing, rinsing and draining fresh produce that you might even consider owning several. Some have a long handle on one side and a hook on the other (to secure the colander to the edge of a saucepan when draining over a pot or bowl); others have small, rounded handles and are best for draining directly into a sink.

Choose one with a large capacity (it is far better to have one that is too large than too small), and a surface that won't easily dent, pit, stain or react with other ingredients. For all these reasons good-quality stainless steel ones are the best and most durable. Enamelled steel colanders can chip, which is very unhygienic, while copper ones can be reactive.

CORER
To neatly remove the cores from apples and pears, an apple corer is indispensible. With a paring knife the result isn't as tidy and there's a risk you could cut all the way through the fruit — or yourself for that matter. A corer needs to be really strong as the pressures exerted on it are great. Get one with the widest head you can find, as some aren't wide enough to remove all the core and seeds. A comfortable rubber handle makes the operation easier as it is less likely to slip in your hand.

FOOD MILL
In this age of labour-saving electrical devices, an old-fashioned manual food mill still has its place. Its ability to separate out skins and seeds while puréeing sauces and soups is a huge advantage over the food processor. Food mills come with a selection of discs — fine, medium and coarse — so you can control the final consistency. You simply rest the unit on top of a bowl or saucepan (with the help of a hook) and then turn a handle, forcing the food under a large blade and then through the chosen disc. When solids under the blade start to slow down the process, you just turn the handle in the reverse direction to dislodge and discard them.

Food mills are either plastic or metal. Stainless steel are best as others can react with food acids and discolour or rust. They come in a range of sizes too — get a large one so you can purée decent quantities at a time. Also check how far the side hook extends — this will determine how large a bowl or saucepan you can purée into.

GARLIC PRESS
Some cooks swear by these for their labour-saving advantages; others loathe them. Garlic presses (or 'crushers' as they are also called) tend to make crushed garlic taste more pungent as they break up more of the potent garlic cells than cutting can usually achieve, and some find the result overpowering.

Choose one with very comfortable handles, a strong construction that can stand up to the rigours of repeated use, and a non-stick, easy-clean surface. Some models claim to press unpeeled cloves, but it's best to peel the garlic before crushing — you get a better yield of garlic and don't have to work so hard for it either. Some presses come with very useful cleaning attachments that flush out embedded bits of garlic from the holes.

GRATER

Whether grating carrots for a carrot cake or potatoes for rösti, a hand grater is a must-have. The most common is a box-style grater, with different textured graters on each side: some for shredding, some for very finely grating and others for coarse grating. Then there are the microplane graters — thin, flat, long stainless steel graters with tiny, extremely sharp blades. These make short work of zesting citrus and grating hard cheeses over pasta dishes. There are also smaller microplane graters for grating spices such as cinnamon and nutmeg for fruit-based cakes and desserts.

IMMERSION BLENDER

Even if you have a food processor, these gizmos are so handy you'll wonder how you managed without one. They look like an electric wand, with a small motor at one end and a rotating blade at the other. With one of these you can purée a batch of soup directly in a saucepan without the mess of transferring it to a food processor.

The best come with a number of attachments that can perform all the functions of a larger food processor. One of the most useful is a mini food processor — brilliant for making a small quantity of pesto, puréeing soups, emulsifying sauces, making pastes, grinding spices, chopping herbs and garlic, whisking and creaming. Choose one with a powerful motor as it will last longer and handle tougher tasks, and also one with all the attachments you think you'll need, including the mini food processor and a whisk.

KNIVES

For vegetable cookery, you need a good paring knife and at least one large or medium-sized cook's knife. A good serrated knife is also very handy for cutting tomatoes, citrus and other tender-fleshed or very juicy produce.

A paring knife is a small knife with a rigid, tapered blade. It makes easy work of removing 'eyes' from potatoes, trimming small fruits and vegetables, peeling the skin from blanched tomatoes and myriad other jobs.

A cook's knife is mainly a chopping knife; the blade can be 15–35 cm (6–14 inches) long. A knife with a 23–25 cm (9–10 inch) blade is versatile enough for a wide range of duties: chopping herbs and vegetables, cutting vegetables into fine julienne strips, slicing corn off a cob and cutting hard fruits like quinces into neat wedges. If your budget can run to two cook's knives, then a large one and a smaller one will cover all the bases.

It is worth spending as much as you can afford on best-quality knives. Choose ones that feel well balanced and very comfortable in your hand. You can tell a good knife by its construction and materials. Knife blades are made of steel, and the best ones tend to be high-carbon stainless steel, which can hold an edge well, don't stain, rust or pit, and can bend without breaking. Look for knives that have a one-piece construction from the tip of the blade to the end of the handle; blades that are riveted or otherwise joined to the handle can come loose over time and are never as strong. Plastic, rubber or stainless steel handles are the most durable. Any rivets or joins to the blade should be completely smooth and seamless so they can't come loose or get food trapped in them.

The tang of a knife is the part of the blade that extends into the handle — a quality knife will have a tang that goes to the end of, and is the same shape as, the handle. Avoid knives where the tang just extends partially into the handle, or where it extends the full length of the handle but is not the same width — these are the hallmarks of cheaper knives and they will not perform well.

MANDOLINE

A mandoline is essentially a type of knife, but one where you move the food over the blade, instead of the other way around. A mandoline is very efficient at reducing fruits or vegetables into consistently thin, neat slices, julienne strips or ripple-cut pieces. Good mandolines have blades that can be changed, depending on what cutting effect you are after. They can be a chore to clean, but nothing else slices or cuts so elegantly.

Mandolines range from the plastic Japanese ones that are basically a plastic box with an angled blade set into the surface that sits flat on a work surface, to pricey stainless steel professional models with many cutting possibilities and a protective carriage to guard your fingers. These mandolines are quite large and expensive but do give impressive results.

MEZZALUNA

Also called a crescent cutter, a mezzaluna is a double-handled knife with a curved blade much used in Italy for quickly reducing large piles of herbs to neatly chopped mounds, without bruising or damaging the herb. You use a rocking motion with these, making them simple and safe to operate. Beautiful and highly functional, they're nice to have in your kitchen if you can afford one. They often come with their own special chopping board.

POTATO MASHER

When a potato is boiled, its starch granules swell. If you break the cell walls of a boiled potato, a lot of starch is released and the texture becomes gluey. A potato masher minimises cell damage, producing mash that is fluffy and light. A potato masher can be used for many other crushing and puréeing jobs as well.

A potato masher should be very comfortable to use and be of good, strong construction to withstand pressure. A sturdy stainless steel one will be the most durable. Make sure the handles are connected to the masher with strong rivets — some of the very expensive ones are moulded in one piece, which is even better. The base should have plenty of holes for the potato to be worked through for the most refined texture.

POTATO RICER

When a smooth, fluffy potato purée is required — for making gnocchi, for example — a potato ricer is a must. (Cooked potato can't be puréed in a food processor as it becomes gluggy.) A potato ricer is like a big garlic press. It has a large, perforated chamber and a heavy arm. Cooked potato goes into the chamber — then you press hard, using a levered handle, to force the potato through the holes. Ricers can also be used for mashing other root vegetables, as well as starchy ones like broad (fava) beans and peas.

Choose a ricer that you can easily rest on top of a pot or bowl during use, with comfortable handles. Do also choose one made of stainless steel as it will be easy to clean. Some have interchangeable sets of plates with differing sized holes, while others have just one in-built plate.

SALAD SPINNER

Salad greens served raw need meticulous washing, and must then be thoroughly dried or dressings will simply slide off them; wet leaves will also quickly become limp. Leaves are tricky to dry because any vigorous handling (shaking too hard, even patting dry with paper towels) can bruise them.

Enter the salad spinner! These work on centrifugal force: as the leaves spin, the water is forced out and collects in a separate chamber inside the spinner. Salad spinners operate either by a crank handle, a pull-cord or a simple pump-action button. They are very easy to use, although the crank-style spinners require slightly more work. Salad spinners are made of plastic and although bulky in size, are quite light.

Look for a sturdy one with a good spinning action and well-fitting lid. Check the spinning mechanism — some are more stable and quieter than others, while some need bracing to secure them when in use.

SAUCEPANS

Choose saucepans with care. Because many fruits and vegetables are acidic or otherwise react, opt for those made from stainless steel or enamelled cast iron. Acids (and alkalis) can corrode aluminium pans, and aluminium can also interact with ingredients such as tomatoes, producing 'off' flavours and colours. Aluminium, however, is light and conducts heat well, so pans with an aluminium core but stainless steel coating are a good compromise.

Stainless steel conducts heat relatively poorly, so good stainless steel saucepans will have a layer of aluminium or copper on the base — this layer should be thick and coat the entire base for even heat conduction. Enamelled cast-iron saucepans are very heavy, so think carefully before buying them. Their surface though is very non-stick — excellent for making sauces and dishes prone to burning (including chutneys or jams).

Always choose saucepans with tight-fitting lids.

STEAMER

A dedicated steamer, or steam basket at the very least, is a very useful kitchen item. Any saucepan can serve as a steamer; you just need some sort of insert in which to steam your vegetables.

Some saucepans are sold with a tight-fitting perforated upper saucepan; the water boils in the bottom saucepan and the vegetables sit in the upper unit, cooking under tight cover in the steam created. Buy good-quality stainless steel steamers, making sure the steamer unit fits very tightly into the lower pan and that the lid also fits snugly.

A steam basket is a less expensive option. These are made of stainless steel and feature collapsible 'petals' that fold up to enclose vegetables to be steamed. They have small legs so that once standing in the saucepan, they will hold the food just above the water level; a central post acts as a handle so you can easily retrieve the steamer from the saucepan.

STRAINER

Strainers have hooks or 'ears' to help them sit securely over a bowl or pan, a rim that supports the wire mesh of the strainer, and a handle. A fine mesh strainer is used for straining fruit or vegetable juice, and larger strainers for straining stocks, sauces, soups and manually making berry and fruit purées.

Choose metal strainers with strong rims that will support a good deal of weight in the mesh 'bowl', and a heat-resistant handle that is both comfortable to hold and well attached to the rim. The finer the mesh of a sieve, the more fragile it is, so choose one that is appropriate to the task. Make sure the metal won't rust or discolour.

larder shelf

Many of the vegetables in this chapter are so familiar we take them for granted. They're the humble ones we usually have on hand that turn up in so many savoury dishes. Many of them also have incredible keeping qualities, and have sustained people through long, hard months when nothing else was available. Today, their popularity endures.

CARROTS

Cuisines the world over make great use of the carrot: it turns up in soups, stews, stocks, stir-fries, salads, sauces, cakes and even desserts and drinks. It contains more vitamin A than any other vegetable and is one of the few root vegetables that tastes as good raw as it does cooked.

Native to Central Asia, carrots were originally purple, black, yellow or even white. It was the Dutch who bred the carrot to be orange, in the eighteenth century. Modern carrots, available year round, are bred to be uniformly straight and rather fat. Baby carrots, sometimes called 'Dutch' carrots, are sold in bunches with their green, feathery leaves still attached.

Carrots should have smooth, uniform skin free of cracks or blemishes, with no shrivelling or browning at the stem. Very orange carrots have higher quantities of beta-carotene (which is converted to vitamin A) than paler ones. Don't buy carrots with green 'shoulders' as these will be bitter; also avoid very large carrots as they will probably be woody and not very sweet. Carrots will last for several weeks in a loosely sealed plastic bag in the crisper.

Carrots are often boiled or steamed, but become particularly sweet when slowly cooked, covered, in a little butter or olive oil, or oven-roasted in olive oil or butter. They are especially complemented by butter, dill, mint, parsley, cream, sour cream, walnuts and almonds, honey, brown sugar, raisins, orange, ginger, cinnamon, nutmeg, cardamom and poppy seeds.

CELERIAC

Although it's hard to see any resemblance, celeriac is related to the carrot (as is celery). This large, round, rather ugly 'ball' has a wild tangle of roots at one end and thick, green, inedible stalks and leaves at the other, with brown roots and fibres giving it a rather earthy appearance. Celeriac was first cultivated in Europe, where it remains most popular.

Celeriac is mainly roasted or steamed (boiling tends to waterlog it), and makes an excellent mash or creamy soup. In thin strips, it can be deep-fried to great effect, or it can be grated or finely julienned and eaten raw, as in the famed French salad, celeriac remoulade (see page 48).

Look for celeriac with the stems and leaves still attached as they indicate how fresh it is. Fresh celeriac has a green tinge around the top of the bulb. Avoid very large celeriac as their texture tends to be a bit 'woolly'. Select bulbs that feel heavy for their size, are firm and about the size of a baseball. To store them, cut the stalks off, place the bulbs in a loosely sealed plastic bag and they will keep in the crisper for up to 10 days.

To prepare celeriac, trim off all the skin (it is quite thick) using a small sharp knife, as well as all the root material at the base of the bulb. Peel it just before using as it turns brown very quickly — to prevent this happening, slip pieces into a bowl of acidulated water (water containing lemon juice) as you go.

Celeriac teams well with butter, apples (celeriac and apple mash is delicious), cream, mayonnaise, walnuts, lemon, mustard, chives and parsley, pickles, gherkins and cheese (blue, gruyère and cheddar).

GARLIC

A member of the allium family (along with onions, shallots and leeks), garlic has long been prized for its medicinal properties. Full of volatile oils, garlic is pungent when used raw in salad dressings, mayonnaise and marinades, but with gentle, slow cooking, its strength dissipates into a pleasant background flavour that many dishes simply aren't the same without.

There are hundreds of garlic varieties, with differing strengths of flavour. Each tight, rounded bulb contains up to 60 cloves, each 'wrapped' in a thin, papery skin. The cloves should be very firm, with no powdery mould or sprouting green shoots. Their papery skins should be crinkly and dry.

Stored in a dark, cool, well-ventilated place, garlic will keep for several weeks. Prepare just before using, as exposure to air turns it bitter (note that the crushed garlic sold in jars never tastes as good as freshly chopped garlic). To peel garlic, smash each clove on a hard surface, using a large knife turned on its side and the heel of your hand. Chopping is best done on a board with a little sea salt, which helps break down the garlic and speeds the process up.

Garlic contains a volatile compound called allinaise, which is released when cloves are chopped — the finer they are chopped, the more allinaise is released. Some recipes call for huge amounts of garlic (such as the classic chicken with 40 cloves of garlic; see page 26), but these dishes work because the garlic is kept whole, and mellows with cooking.

The sharpness of raw garlic adds a wonderful accent to butter, oil and mayonnaise as well as some salads and vegetable dishes. A little cooked garlic goes with practically any meat or vegetable-based dish.

Onion

Cultivated since ancient times, the onion is surely one of the great kitchen stalwarts. With the exception of some religious communities (such as the Jains and some Buddhist groups), they are used in cooking the globe over.

Available year round, there are about 300 varieties, but most onions we buy are the classic type with the light brown skin that come in a variety of sizes and have quite a strong flavour; these onions are best used in cooking as they are too pungent to eat raw.

Red onions, with their pretty mauve-red skins and red-ringed interiors, are milder and sweeter and are perfect for salads, salsas and wherever raw onion is required. They can of course be cooked, but their flavour gets a little lost in long-cooked dishes. They're great for grilling, roasting in large pieces or barbecuing. **Spanish onions** are large, round, juicy, yellow-skinned onions that have a sweet, mellow flavour and are ideal for using raw. **White onions** have a white skin and greeny-white flesh; these tend to be stronger-tasting than even the brown varieties and are excellent for cooking. **Sweet onions** have a sugar content of about 15%, compared to 3–5% in most onions. Sweet onions have thin, fragile skins. Their mild, subtle flavour makes them perfect for raw use. **Baby** or **pickling onions** are miniature versions of (usually) brown or white onions. They can be pickled whole, but are also great in casseroles and vegetable dishes.

Choose firm onions that are heavy for their size, with no sign of green sprouting. The skin should be shiny and tight and closed around the neck, and feel perfectly dry. Onions shouldn't smell strongly 'oniony' until they have been cut; a strong smell can indicate spoilage. Store onions in a cool, dark place with good ventilation; they will quickly go mouldy in humid conditions. Don't store them near potatoes, as these give off moisture and cause them to spoil. Stored properly, onions will keep for at least 1 month.

Some of the most sublime flavour partners for onion include anchovies, capers, olives, balsamic vinegar, beef, venison, liver (and generally all roasted and barbecued meats), bacon, thyme, basil, dill, sour cream, feta, parmesan, gruyère and blue cheese, raisins, tomato, red wine and rosemary.

Parsnip

Closely related to the carrot, parsnips were the main winter vegetable for most of Europe until the potato came along. Associated with the comforting flavours of winter, the parsnip has a wonderful sweet flavour unlike any of the other root vegetables, and is quite high in vitamin C and potassium.

Parsnips are sold without their green, leafy tops, which contain an irritating substance. Fresh ones have a smooth, evenly creamy skin and should be hard, not bendy. They should feel heavy for their size. Avoid any that are bruised or blemished or have tops that are starting to sprout or look brown; sprouting parsnips will most likely be woody inside. Parsnips will keep for about 1 week in the crisper in a loosely sealed plastic bag.

To prepare parsnips, trim away their stalk end and peel away the thick, bitter peel just before cooking as they turn brown on contact with air. Do not put them into acidulated water to stop them browning as they will absorb water, which will ruin their texture.

Roasting turns the exterior chewy, golden and sweet and the inside fluffy and light — a classic accompaniment to winter roasts and stews. Cooled roasted parsnips are also excellent tossed into salads. Only boil parsnips if you plan to mash or purée them, in which case they will benefit from further cooking over low heat to evaporate any excess water. Parsnip makes excellent soup, particularly with potato (with which it is also lovely mashed with butter and cream). Grated, it can also be used in fritters, cakes and puddings, as one would use carrot.

Parsnip pairs beautifully with cream, butter, nutmeg, walnuts and hazelnuts (and their oils), orange, balsamic vinegar, gruyère and parmesan cheese, sausages, ham, bacon, prosciutto, game birds, beef, chicken and lamb. It is also complemented by parsley, sage, thyme and garlic.

Potatoes

The humble potato is eaten all over the world, providing more energy than any other food crop. A member of the nightshade family and related to the tomato, it is the tuber of a flowering vine and is grown year round.

New potatoes are those harvested when they are still a little immature. They have dense, moist flesh and very thin, fragile skins. They have a particularly 'fresh' potato flavour and are delicious boiled, with perhaps some fresh mint and a slathering of good butter.

Old potatoes, also called 'maincrop' potatoes, are the type most commonly sold. They have drier flesh and thicker skins than new potatoes and keep well stored in cool dark conditions. There are many different types of potatoes, any of which can be harvested when 'new' or 'old'.

While most potatoes have white, creamy or yellowish flesh, some have blue or even purple flesh; these are particularly ancient types and are fun to experiment with. Some have red skin, others golden-brown; some are finger-shaped, while others are tiny and can be cooked whole. Different countries have their own favourite varieties, from the toolangi delights and colibans of Australia to the yukon gold, concord and butterball of America to England's maris piper and pink fir apple.

The most important thing is to find out which types are waxy and which ones are floury, as this determines how they are best cooked. Some potatoes, however, are 'all purpose', which means they can be cooked any way you wish. Floury potatoes (this includes most old potatoes) have more starch than new potatoes. They make fantastically light fluffy mash and are good for baking in their jackets and roasting too, but are not ideal for making chips or to use in salads. Waxy potatoes (new potatoes and most red varieties) have firmer flesh with less starch. They are great for boiling whole, for use in salads and for making chips. They can also be roasted, but won't be as light and fluffy inside as old potatoes. They're not suitable for mashing as the end result will be wet and gluey.

Buy potatoes loose, not in plastic bags, so you can assess them individually. Avoid any with greening skins (the greening, which goes through the flesh too, is toxic). They should feel firm and heavy, with no bruises or blemishes. Avoid any with 'eyes' that are starting to sprout.

With the exception of new potatoes, which do not store well at all and should be used within a few days of purchase, store potatoes in a cool, dark, dry place in a cloth sack or heavy paper bag (never refrigerate them). Stored correctly, they will keep for up to 3 months. During this time, their starches slowly convert to sugar, so older potatoes will be sweeter.

When cooking potatoes, remember that most of their valuable nutrients (including vitamins B6 and C, niacin, iodine, folic acid, copper and magnesium) are concentrated just under their skin. Wash them well but don't peel them unless you really need to for aesthetic reasons (such as when making gnocchi, for example).

PUMPKIN (WINTER SQUASH)

Native to Central America, the versatile pumpkin is the fruit of a rambling vine, used in myriad dishes savoury and sweet. Its name derives from a Greek word, *pepon*, meaning 'large melon', and pumpkins can indeed be huge — the largest one on record weighed about 767 kg (1689 lb)!

There are hundreds of different types, varying greatly in size and shape. Some are bell shaped, some round, others squat. The skin can be smooth, ridged or bumpy, and vary in colour from greyish white to dark olive green. Their flesh colour, sweetness, water content and texture can all vary wildly, but they are basically interchangeable for cooking purposes.

Larger pumpkins are often sold cut into pieces, wrapped in plastic. When buying cut pieces, choose ones with moist, dense, darkly coloured flesh. Cut pumpkin should be stored in the refrigerator, in a loosely sealed plastic bag, for only 2–3 days. An uncut pumpkin in good condition will keep for several months in a cool, dark, well-ventilated place.

All pumpkins need to have their seeds removed before cooking; most also require peeling before use, although some thinner-skinned varieties can be roasted with the skin on. Pumpkin is better steamed than boiled, as boiling can make it watery. Chopped pumpkin can also be cooked, covered, in just a little butter or oil over low–medium heat; as with roasting, the flavour becomes deliciously concentrated.

Pumpkin marries well with all manner of cheeses, cream, eggs, butter, bacon and other salty pork meats, roast meats, rocket (arugula), watercress, spinach, sweet spices, tomatoes, olives, orange and olives. It is excellent with many herbs, including sage, rosemary and marjoram, and sweet flavours such as maple syrup, brown sugar, cinnamon and honey.

SHALLOTS

Related to onions and garlic, but with a milder flavour, these small, elongated, onion-like bulbs have thin, papery, pink, yellow or coppery brown skins, and can be as small as a tiny fingernail or nearly the size of a pickling onion. (Shallots are not to be confused with spring onions or scallions, which have a slender white root end and long green leaves.) The small pinkish shallots are considered the finest-flavoured.

Shallots are at their peak in summer. Choose ones that feel very firm and have bright, brittle skins. Shallots in good condition will keep in a paper or string bag in a cool, dark, well-ventilated place for several months.

To make the skins easier to remove, soak the bulbs in boiling water for 5 minutes. Drain well and use a small sharp knife to help pull the skins away from the bulb, down towards the root end. Trim the root ends, but don't cut them off completely if you wish to cook the shallots whole or in halves (the root end will help hold them together).

Shallots are widely used in French and Thai cookery in many savoury dishes. They can also be braised or roasted whole.

Sweet potato

Available year round, sweet potatoes are not related to potatoes at all. Native to Central and South America, they were spread by the Spanish to Europe, and as far afield as Asia and Africa, where they have become an important staple. There are many varieties, some of which are erroneously called 'yams', and their size varies hugely. Their skin colour ranges from deepest purple through to orange or yellow-beige. Their sweetness and flesh colour also vary; some cook to be dry and fluffy, others rather moist.

Look for ones with smooth, unblemished skins and store them in a cool, dark, well-ventilated place (never the fridge) for no more than a week. They can be cooked in much the same way as potato. They make delicious mash, fritters and salads and are great roasted, steamed, pan-fried or baked.

They are also delicious in pies and cakes and are wonderful with maple syrup, brown sugar, honey, sweet spices (cinnamon, ginger, cloves, allspice, cardamom), saffron, roast meats, caramelised onions, bacon and ham, sage, rosemary and coriander (cilantro).

Turnips and swedes (rutabaga)

Turnips were possibly one of the first crops cultivated. The swede is much more recent — the result of a seventeenth-century crossing of a turnip with a cabbage, hence its mustardy flavour. (They are supposedly called swedes because they are such a common crop in Sweden!)

Both of these root vegetables are round, with long stems and darkish green leaves. Turnips have creamy white flesh and a whitish skin that deepens to a pretty red-purple around the top. Swedes are a light orangey colour, changing to a muted greeny-mauve around the top. Both have thick skin that needs to be pared away entirely using a small sharp knife.

Look for turnips with tight, clear skin. Swedes should also have unblemished skin and feel heavy for their size. Turnips will keep in a cool, well-ventilated place for up to 1 week, and swedes for 1 month.

Both vegetables are better steamed than boiled as boiling turns them soggy. They make great mash (on their own, or with potato, parsnip or carrot), add a sweet accent to meaty braises and vegetable soups, and are delicious in creamy gratins and fritters. They go well with cream, sour cream, butter, apple, pear, dill, chives, parsley, duck, lamb, beef, cured meats, cheddar and gruyère cheese.

PRAWNS WITH SAFFRON POTATOES

SERVES 4

16 raw prawns (shrimp)
4 tablespoons olive oil
450 g (1 lb) new potatoes, cut in half
$1/4$ teaspoon saffron threads, lightly
   toasted and then crushed
1 garlic clove, crushed
1 red chilli, seeded and finely chopped
1 teaspoon grated lime or lemon zest
3 tablespoons lime juice
200 g (7 oz) baby rocket (arugula)

Preheat the oven to 180°C (350°F/Gas 4). Peel and remove the veins from the prawns, leaving the tails intact. Set aside.

Heat 2 tablespoons of the olive oil in a large frying pan. Add the potatoes and cook over medium heat, without turning, for 6–7 minutes, or until golden. Transfer to a roasting tin and toss with the saffron and some sea salt and freshly ground black pepper. Bake for 25 minutes, or until tender.

Heat a chargrill pan or barbecue hotplate to medium. Toss the prawns in a bowl with the garlic, chilli, citrus zest and 1 tablespoon of the remaining olive oil. Grill the prawns for 2 minutes on each side, or until pink and just cooked through.

Put the lime juice and remaining olive oil in a small screwtop jar and shake together well. Season with sea salt and freshly ground black pepper.

Divide the potatoes, rocket and prawns among four serving plates and drizzle with the dressing.

CHICKEN WITH FORTY CLOVES OF GARLIC
SERVES 4

10 g (¼ oz) unsalted butter
1 tablespoon olive oil
2 kg (4 lb 8 oz) whole free-range chicken
40 garlic cloves, unpeeled (see Note)
2 tablespoons chopped rosemary
2 thyme sprigs
270 ml (9½ fl oz) dry white wine
150 ml (5 fl oz) chicken stock
225 g (8 oz/heaped 1¾ cups) plain
  (all-purpose) flour

Preheat the oven to 180°C (350°F/Gas 4).

Melt the butter with the olive oil in a 4.5 litre (157 fl oz/ 18 cup) flameproof casserole dish with a tight-fitting lid. Add the whole chicken and fry over medium heat for about 8 minutes, turning often to brown all over. Remove the chicken from the dish.

Add the garlic cloves, rosemary and thyme to the dish and cook for 1 minute. Return the chicken to the dish and pour in the wine and stock. Bring to a simmer, basting the chicken with the liquid.

Put the flour in a bowl and add 150 ml (5 fl oz) water, stirring to form a firm paste. Divide the mixture into four even portions and roll each one into a cylinder. Place the dough cylinders around the rim of the casserole dish, joining the ends together, then cover with the lid, pressing down well to seal. Bake for 1¼ hours.

Remove the lid by carefully cracking the paste, then bake the chicken for a further 15 minutes to brown it. Transfer the chicken to a warmed plate, reserving the cooking juices, then cover loosely with foil and leave to rest.

Meanwhile, bring the cooking juices to the boil, then simmer over medium heat for 7–10 minutes, or until reduced to about 250 ml (9 fl oz/1 cup).

Carve the chicken, then pierce the garlic skins and squeeze the garlic flesh onto the chicken. Serve with the cooking juices spooned over.

NOTE: Don't be alarmed by the amount of garlic in this recipe. The garlic becomes soft, creamy and mildly flavoured when cooked in this way.

CREAMY GARLIC SEAFOOD

SERVES 6

6 raw moreton bay bugs (flat-head
  lobsters)
500 g (1 lb 2 oz) raw prawns (shrimp)
50 g (1¾ oz) unsalted butter
1 onion, finely chopped
5–6 large garlic cloves, finely chopped
125 ml (4 fl oz/½ cup) white wine
500 ml (17 fl oz/2 cups) pouring
  (whipping) cream
1½ tablespoons dijon mustard
2 teaspoons lemon juice
500 g (1 lb 2 oz) skinless perch fillets,
  cut into bite-sized cubes (see Note)
12 scallops, with roe, cleaned
2 tablespoons chopped flat-leaf (Italian)
  parsley

Cut the heads off the bugs, then use kitchen scissors to cut through the shell around each side of the tail so you can open the shell. Remove the tail flesh in one piece.

Peel and remove the veins from the prawns, leaving the tails intact, then set aside.

Melt the butter in a frying pan and sauté the onion and garlic over medium heat for 2 minutes, or until the onion has softened. Add the wine and cook for 4 minutes, or until reduced by half. Stir in the cream, mustard and lemon juice and simmer for 5–6 minutes, or until the liquid has reduced to almost half.

Add the prawns and cook for 1 minute, then add the bug meat and cook for another minute, or until opaque. Add the fish and cook for 2 minutes, or until cooked through (the flesh will flake easily when tested with a fork). Finally, add the scallops and cook for 1 minute, or until all the seafood is cooked.

Remove the pan from the heat and toss the parsley through. Season to taste with sea salt and freshly ground black pepper. Serve with salad and bread.

VARIATION: Instead of perch you can use any firm, white-fleshed fish such as ling, bream, snapper or blue-eye fillets.

POTATO AND SAGE CHIPS

SERVES 4–6 AS A SNACK OR TO SERVE WITH DRINKS

2 large all-purpose potatoes
2 tablespoons olive oil
25 sage leaves
sea salt, for sprinkling

Preheat the oven to 200°C (400°F/Gas 6). Using a large sharp knife or a mandoline, cut the potatoes lengthways to give 50 paper-thin slices. Toss the slices in the olive oil.

Line two baking trays with baking paper. Sandwich a sage leaf between two slices of potato and sprinkle with sea salt. Repeat with the remaining potatoes and sage leaves.

Arrange on the baking trays in a single layer and bake for 25–30 minutes, or until the chips are browned and crisp, turning once — watch carefully during cooking so they don't burn (some chips may cook slightly more quickly than others). Serve hot.

SWEET POTATO ROSTI

MAKES ABOUT 30

2 orange sweet potatoes (about 800 g/
   1 lb 12 oz in total), unpeeled
2 teaspoons cornflour (cornstarch)
1/2 teaspoon sea salt
40 g (1 1/2 oz) unsalted butter
150 g (5 1/2 oz) mozzarella cheese,
   cut into 30 pieces

Boil the sweet potatoes for 15 minutes, or until almost cooked, but still firm. Set aside to cool, then peel and roughly grate into a bowl. Add the cornflour and sea salt and toss lightly to combine.

Meanwhile, preheat the oven to 120°C (235°F/Gas 1/2).

Melt some of the butter in a non-stick frying pan over medium heat. Spoon teaspoons of the sweet potato mixture into the pan and put a piece of cheese in the centre of each. Top with another teaspoon of the sweet potato mixture and gently flatten to form rough circles. Cook for 3 minutes on each side, or until golden.

Remove with a slotted spoon and drain on paper towels, then place in the oven to keep warm while cooking the remaining rösti. Serve hot.

NOTE: The sweet potato can be cooked and grated up to 2 hours ahead and set aside, covered, until ready to cook.

Potato and sage chips

SWEET POTATO, PUMPKIN AND COCONUT PIES
MAKES 8

2 tablespoons vegetable oil
1 onion, finely chopped
2 garlic cloves, crushed
1 teaspoon grated fresh ginger
1 small red chilli, chopped
1 small orange sweet potato
　(250 g/9 oz), peeled and chopped
250 g (9 oz) pumpkin (winter squash),
　peeled and chopped
1/2 teaspoon fennel seeds
1/2 teaspoon yellow mustard seeds
1/2 teaspoon ground turmeric
1/2 teaspoon ground cumin
150 ml (5 fl oz) coconut milk
3 tablespoons chopped coriander
　(cilantro) leaves
4 sheets of frozen puff pastry, thawed
1 egg yolk

Heat the oil in a heavy-based saucepan. Add the onion, garlic, ginger and chilli and sauté over medium heat for 5 minutes, or until the onion has softened. Add the sweet potato, pumpkin, fennel and mustard seeds, turmeric and cumin. Stir for 2 minutes, then stir in the coconut milk and 2 tablespoons water. Reduce the heat to low and simmer for 20 minutes, or until the vegetables are tender, stirring frequently. Stir in the coriander and set aside to cool.

Grease a large baking tray. Cut eight 9 cm (3 1/2 inch) circles from two sheets of the pastry and place them on the tray. Divide the filling among the circles, mounding the filling slightly in the centre, and spreading to within 1 cm (1/2 inch) of the edge. Brush the edges of the pastry with a little water.

Using a lattice cutter or a sharp knife, cut out eight 10 cm (4 inch) circles from the remaining pastry. Carefully fit them over the filling, then press the edges together firmly to seal. Using the back of a knife, press the outside edge lightly at 1 cm (1/2 inch) intervals. Refrigerate for at least 20 minutes.

Meanwhile, preheat the oven to 190°C (375°F/Gas 5).

Mix the egg yolk with 1 teaspoon water and brush the egg glaze over the pastry. Bake for 20–25 minutes, or until golden. Serve warm.

PUMPKIN AND BORLOTTI BEAN SOUP
SERVES 4–6

350 g (12 oz) dried borlotti (cranberry)
   beans
1 kg (2 lb 4 oz) butternut pumpkin
   (squash), peeled, seeded and chopped
2 large boiling potatoes, such as long
   white, peeled and chopped
2 litres (70 fl oz/8 cups) chicken stock
1 tablespoon olive oil
1 red onion, chopped
2 garlic cloves, finely chopped
1 celery stalk, sliced
6–8 sage leaves, chopped
crusty Italian bread, to serve

Put the beans in a large bowl, cover with plenty of cold water and leave to soak overnight.

Rinse the beans well, then place in a saucepan, cover with plenty of fresh cold water and bring to the boil. Reduce the heat and simmer for 1 1/2 hours, or until tender. Drain well and set aside.

Put the pumpkin, potato and stock in a large saucepan. Bring to the boil, then reduce the heat and simmer for 35–40 minutes, or until the vegetables are soft. Drain well, reserving the liquid. Mash the pumpkin and potatoes, then return to the pan with the reserved liquid. Stir in the beans.

Heat the olive oil in a saucepan. Add the onion, garlic and celery and sauté over medium heat for 6–7 minutes, or until softened. Add to the soup with the sage, season with freshly ground black pepper and gently heat through.

Serve hot, with crusty Italian bread.

Sweet potato ravioli

Serves 4

500 g (1 lb 2 oz) orange sweet potato, peeled and cut into 2 cm (³/4 inch) chunks
3 tablespoons olive oil
150 g (5½ oz/²/3 cup) ricotta cheese
2 tablespoons grated parmesan cheese
1 tablespoon chopped basil
3 garlic cloves, crushed
500 g (1 lb 2 oz) egg won ton wrappers
60 g (2¼ oz) unsalted butter
310 ml (10¾ fl oz/1¼ cups) pouring (whipping) cream
small basil leaves, to garnish

Preheat the oven to 220°C (425°F/Gas 7). Place the sweet potato on a baking tray and drizzle with the olive oil. Bake for 40–45 minutes, or until tender.

Transfer the sweet potato to a bowl. Add the ricotta, parmesan, basil and one-third of the garlic and mash until smooth. Season to taste with sea salt and freshly ground black pepper.

Line a baking tray with baking paper. Cover the won ton wrappers with a damp tea towel (dish towel) to stop them drying out. Place 2 level teaspoons of the sweet potato mixture into the centre of one won ton wrapper and brush the edges with a little water. Top with another wrapper, pressing the edges together to seal. Place on the baking tray and cover with a tea towel. Repeat with the remaining ingredients to make 24 ravioli, placing a sheet of baking paper between each layer on the tray.

Melt the butter in a frying pan. Add the remaining garlic and sauté over medium heat for 1 minute. Add the cream, bring to the boil, then reduce the heat and simmer for 4–5 minutes, or until the cream has reduced and thickened. Cover and keep warm.

Bring a large saucepan of water to the boil. Cook the ravioli in batches for 2–4 minutes, or until just tender. Drain well and divide among warmed serving bowls. Ladle the hot sauce over the top, scatter with basil leaves and serve.

SPICED PUMPKIN AND LENTIL TAGINE
SERVES 4–6

275 g (9³/4 oz/1¹/2 cups) brown lentils,
    rinsed
2 roma (plum) tomatoes
600 g (1 lb 5 oz) firm pumpkin
    (winter squash) or butternut pumpkin
    (squash)
3 tablespoons olive oil
1 onion, finely chopped
3 garlic cloves, finely chopped
¹/2 teaspoon ground cumin
¹/2 teaspoon ground turmeric
a pinch of cayenne pepper
1 teaspoon paprika
3 teaspoons tomato paste (concentrated
    purée)
¹/2 teaspoon sugar
1 tablespoon finely chopped flat-leaf
    (Italian) parsley
2 tablespoons chopped coriander
    (cilantro) leaves
pitta bread, to serve

Put the lentils in a saucepan with 1 litre (35 fl oz/4 cups) cold water. Bring to the boil, skimming off any impurities that rise to the surface. Reduce the heat, then cover and simmer for 20 minutes, or until nearly tender. Drain well.

Meanwhile, cut the tomatoes in half widthways and squeeze out the seeds. Coarsely grate the tomatoes into a bowl, down to the skin, discarding the skin. Set aside.

Peel the pumpkin, remove the seeds, then cut the flesh into 3 cm (1¹/4 inch) cubes. Set aside.

Heat the olive oil in a large saucepan. Add the onion and sauté over low heat for 6–7 minutes, or until softened. Add the garlic, cook for a few seconds, then stir in the cumin, turmeric and cayenne pepper. Cook for 30 seconds, or until fragrant.

Stir in the grated tomatoes, paprika, tomato paste, sugar, half the parsley and half the coriander. Season with sea salt and freshly ground black pepper.

Add the lentils and pumpkin, stir well, then cover and simmer for 20 minutes, or until the pumpkin and lentils are very tender. Adjust the seasoning and sprinkle with the remaining parsley and coriander. Serve hot or warm, with pitta bread.

LAYERED POTATO AND APPLE BAKE

SERVES 6 AS A SIDE DISH

2 large potatoes
3 granny smith apples
1 onion
60 g (2¼ oz/½ cup) finely grated
   cheddar cheese
250 ml (9 fl oz/1 cup) pouring
   (whipping) cream
¼ teaspoon ground nutmeg

Preheat the oven to 180°C (350°F/Gas 4). Grease a large, shallow baking dish.

Peel the potatoes and thinly slice. Peel, halve and core the apples, then thinly slice. Peel the onion, then slice into very fine rings.

Layer the potato, apple and onion in the baking dish, finishing with a layer of potato. Sprinkle with the cheese, then pour the cream over. Sprinkle with the nutmeg and some freshly ground black pepper.

Bake for 45 minutes, or until the top is golden brown and the potato is tender. Remove from the oven and allow to stand for 5 minutes before serving.

NOTE: To stop the potato and apple slices browning before assembling the dish, put them in a bowl of cold water with a good squeeze of lemon juice. Drain and pat dry with paper towels just before using.

FRENCH SHALLOT TATIN
SERVES 6

750 g (1 lb 10 oz) large brown shallots, unpeeled
50 g (1 3/4 oz) unsalted butter, plus extra, for greasing
2 tablespoons olive oil
4 tablespoons soft brown sugar
3 tablespoons balsamic vinegar

PASTRY
125 g (4 1/2 oz/1 cup) plain (all-purpose) flour
60 g (2 1/4 oz) cold unsalted butter, chopped
2 teaspoons wholegrain mustard
1 egg yolk, mixed with 1 tablespoon iced water

Put the shallots in a saucepan of boiling water for 5 minutes to make them easier to peel. Drain well, allow to cool slightly, then peel the shallots, taking care to leave the root ends intact.

Heat the butter and olive oil in a large heavy-based frying pan. Add the shallots and cook over low heat, stirring often, for 15 minutes, or until the shallots have started to soften. Add the sugar, vinegar and 3 tablespoons water and stir to dissolve the sugar. Simmer over low heat for a further 15–20 minutes, or until the liquid has reduced and has become syrupy, turning the shallots occasionally.

To make the pastry, sift the flour and a pinch of sea salt into a large bowl. Using your fingertips, lightly rub the butter and mustard into the flour until the mixture resembles coarse breadcrumbs. Make a well in the centre. Add the egg yolk mixture to the well and mix using a flat-bladed knife until a dough forms. Gently gather the dough together, transfer to a lightly floured surface, then press into a round disc. Cover with plastic wrap and refrigerate for 30 minutes, or until firm.

Meanwhile, preheat the oven to 200°C (400°F/Gas 6).

Grease a shallow 20 cm (8 inch) round sandwich tin with butter. Pack the shallots tightly into the tin and drizzle with any syrup remaining in the frying pan.

On a sheet of baking paper, roll out the pastry to a circle 1 cm (1/2 inch) larger than the sandwich tin. Lift the pastry into the tin and lightly push it down so it is slightly moulded over the shallots. Bake for 20–25 minutes, or until the pastry is golden brown.

Remove the tart from the oven and allow to stand in the tin on a wire rack for 5 minutes. To serve, place a plate over the tin and carefully turn the tart out, then gently invert the tart onto a serving plate. Serve warm.

Shallot tatin is best eaten the day it is made.

ONION TART
SERVES 6

PASTRY
250 g (9 oz/2 cups) plain (all-purpose) flour
150 g (5½ oz) cold unsalted butter, chopped
75 ml (2¼ fl oz) iced water

50 g (1¾ oz) unsalted butter
550 g (1 lb 4 oz) onions, finely sliced
2 teaspoons thyme leaves
3 eggs
270 ml (9½ fl oz) thick (double/heavy) cream
60 g (2¼ oz/½ cup) grated gruyère cheese
freshly grated nutmeg, to taste

To make the pastry, sift the flour and a pinch of sea salt into a large bowl. Using your fingertips, lightly rub the butter into the flour until the mixture resembles coarse breadcrumbs. Make a well in the centre. Add the iced water to the well and mix using a flat-bladed knife until a rough dough forms. Gently gather the dough together, transfer to a lightly floured surface, then press into a round disc. Cover with plastic wrap and refrigerate for 30 minutes.

Meanwhile, preheat the oven to 180°C (350°F/Gas 4).

On a lightly floured surface, roll out the pastry to a circle large enough to cover the base and side of a 23 cm (9 inch) fluted, loose-based tart tin. Ease the pastry into the tin, then line the pastry shell with baking paper and half-fill with baking beads, dried beans or rice.

Bake the pastry for 10 minutes, then remove the paper and baking beads and bake for a further 3–5 minutes, or until the pastry is just cooked and dry to touch.

Meanwhile, melt the butter in a small heavy-based frying pan. Add the onion and sauté over low heat for 10–15 minutes, or until soft and golden. Add the thyme and stir well, then set aside and leave to cool.

Whisk together the eggs and cream, then stir in the cheese. Season with sea salt, freshly ground black pepper and a little grated nutmeg.

Spoon the onion into the pastry shell and smooth the surface even. Pour the egg mixture over and bake for 35–40 minutes, or until golden brown and set in the middle.

Remove from the oven and leave to stand in the tin for 5 minutes before slicing. Serve hot or warm.

Onion tart is best eaten the day it is made.

SWEET AND SOUR ONIONS
SERVES 6 AS A SIDE DISH

3 red onions (about 500 g/1 lb 2 oz)
2 tablespoons wholegrain mustard
2 tablespoons honey
2 tablespoons red wine vinegar
2 tablespoons olive oil

Preheat the oven to 220°C (425°F/Gas 7).

Carefully peel the onions, keeping the root end intact so that the layers stay together. Cut each onion lengthways into eight pieces and place in a non-stick baking dish.

Whisk together the remaining ingredients, then drizzle over the onions. Cover and bake for 20 minutes, then remove the cover and bake for a further 15–20 minutes, or until the onion is soft and caramelised. Serve hot.

FRENCH ONION SOUP
SERVES 6

60 g (2¼ oz) unsalted butter
700 g (1 lb 9 oz) onions, finely sliced
2 garlic cloves, finely chopped
4 tablespoons plain (all-purpose) flour
2 litres (70 fl oz/8 cups) beef or chicken stock
250 ml (9 fl oz/1 cup) white wine
1 bay leaf
2 thyme sprigs
12 slices of day-old baguette
115 g (4 oz/heaped ¾ cup) grated gruyère cheese

Melt the butter in a heavy-based saucepan over low heat. Add the onion and cook, stirring occasionally, for 25 minutes, or until deep golden brown.

Add the garlic and flour and cook, stirring often, for 2 minutes. Gradually pour in the stock and wine, stirring constantly to prevent lumps forming. Bring to the boil and add the bay leaf and thyme. Reduce the heat, cover and simmer for 25 minutes, then remove the bay leaf and thyme sprigs and season with sea salt and freshly ground black pepper.

Preheat the oven grill (broiler) to high. Place the baguette slices on a tray, then toast under the grill on both sides. Divide the toasts among six warmed shallow flameproof soup bowls and ladle the soup over the top. Sprinkle the cheese over the soup and heat under the grill until the cheese melts and turns a light golden brown. Serve immediately.

CARROT AND PUMPKIN RISOTTO
SERVES 6

90 g (3¼ oz) unsalted butter
1 onion, finely chopped
400 g (14 oz) pumpkin (winter squash),
   peeled, seeded and finely chopped
   to give 300 g (10½ oz/2 cups)
3 carrots, chopped
2 litres (70 fl oz/8 cups) vegetable or
   chicken stock, approximately
440 g (15½ oz/2 cups) risotto rice
90 g (3¼ oz/1 cup) shaved pecorino
   or parmesan cheese
¼ teaspoon freshly grated nutmeg
½ teaspoon thyme leaves

Heat 60 g (2¼ oz) of the butter in a large heavy-based saucepan. Add the onion and sauté over medium heat for 2–3 minutes, or until beginning to soften.

Add the pumpkin and carrot and cook for 6–8 minutes, or until tender. Mash the mixture slightly using a potato masher.

Meanwhile, pour the stock into a separate saucepan and bring to the boil. Reduce the heat, then cover and keep at simmering point.

Add the rice to the vegetables and cook for 1–2 minutes, stirring constantly, until the grains are translucent and heated through. Add 125 ml (4 fl oz/½ cup) of the simmering stock and cook, stirring constantly, until all the stock has been absorbed. Continue adding the stock, 125 ml (4 fl oz/½ cup) at a time, stirring constantly and making sure the stock has been absorbed before adding more. Cook for 20–25 minutes, or until the rice is tender and creamy; you may need slightly less or more stock.

Remove from the heat, then add the remaining butter, cheese, nutmeg and thyme. Season with freshly ground black pepper and stir thoroughly. Cover and leave for 5 minutes before serving.

TUNISIAN CARROT SALAD
SERVES 6 AS A SIDE DISH

500 g (1 lb 2 oz) carrots, thinly sliced
3 tablespoons finely chopped flat-leaf
  (Italian) parsley
1 teaspoon ground cumin
4 tablespoons olive oil
3 tablespoons red wine vinegar
2 garlic cloves, crushed
¼–½ teaspoon harissa (see Note)
2–3 teaspoons honey, or to taste
  (optional)
12 black olives
2 hard-boiled eggs, peeled and quartered

Cook the carrot in boiling salted water until tender. Drain well and transfer to a bowl.

Add the parsley, cumin, olive oil, vinegar, garlic and harissa to taste, if using. Season to taste with sea salt and freshly ground black pepper. If your carrots are not particularly sweet, drizzle with a little honey, then gently stir together.

To serve, place the carrot mixture in a serving dish, scatter the olives over the top and garnish with the egg quarters. Serve at room temperature.

NOTE: Harissa is a spicy, fiery chilli paste widely used in North African cookery. You can buy harissa from delicatessens, or easily make it yourself. The following recipe fills a 600 ml (21 fl oz) jar and will keep in the fridge for up to 6 months. It is delicious with tagines and couscous, or can be added to soups, salad dressings, marinades, pasta sauces, casseroles and dried bean salads.

To make harissa, remove the stems from 125 g (4½ oz) dried red chillies. Roughly chop the chillies and soak in boiling water for 1 hour. Drain, place in a food processor and add 1 tablespoon dried mint, 1 tablespoon ground coriander, 1 tablespoon ground cumin, 1 teaspoon ground caraway seeds, ½ teaspoon sea salt, 10 chopped garlic cloves and 1 tablespoon olive oil. Process for 20 seconds, scrape down the bowl, then process for another 30 seconds. Add another 2 tablespoons olive oil and process again. Add a further 2–3 tablespoons olive oil and process until a thick paste forms. Spoon the paste into a hot, sterilised jar (for instructions on sterilising jars, see Note on page 69), cover with a thin layer of olive oil and seal. Label and date the jar and store in the fridge.

DUCK WITH TURNIPS

SERVES 2

1 bouquet garni
1.8 kg (4 lb) whole duck
30 g (1 oz) clarified butter
1 carrot, chopped
1 celery stalk, chopped
½ large onion, chopped
2 teaspoons sugar
8 shallots, peeled
8 baby turnips, peeled
100 ml (3½ fl oz) white wine
500 ml (17 fl oz/2 cups) chicken stock
10 g (¼ oz) butter, softened
2 teaspoons plain (all-purpose) flour

Preheat the oven to 180°C (350°F/Gas 4) and put a roasting tin in the oven to heat.

Put the bouquet garni in the duck cavity, then tie the legs together with kitchen string and tie the wing tips together behind the body. Prick the skin all over with a metal skewer.

Melt the clarified butter in a large heavy-based frying pan. Add the duck and brown all over. Lift the duck out of the pan and pour all but 1 tablespoon of the fat into a pouring jug.

Add the carrot, celery and onion to the pan and sauté over medium heat for 5–6 minutes, or until softened. Remove the vegetables and set aside.

Add another 2 tablespoons duck fat to the pan. Add the sugar and let it dissolve over low heat. Increase the heat to high, add the shallots and turnips and cook, turning often, for several minutes, or until golden. Remove from the pan and set aside.

Pour the wine into the pan and boil for 30 seconds, stirring to loosen any bits stuck to the bottom of the pan.

Arrange the carrot, celery and onion in the middle of the hot roasting tin, put the duck on top and pour in the wine and stock. Roast for 45 minutes.

Baste the duck well, add the turnips and shallots and roast for a further 20 minutes. Baste again and roast for a final 25 minutes.

Lift out the duck, turnips and shallots and keep warm. Strain the sauce, pressing down on the vegetables in the sieve to extract all the juices, then discard the vegetables.

Pour the strained sauce into a saucepan and boil rapidly until reduced by half. Mix together the butter and flour, then whisk into the sauce. Boil, stirring constantly to prevent lumps forming, for 2 minutes, or until the sauce has thickened.

Arrange the duck, turnips and shallots on a serving plate and pour a little sauce over them. Serve the remaining sauce separately for pouring over.

FINNISH CREAMY BAKED SWEDE

SERVES 6–8 AS A SIDE DISH

1.6 kg (3 lb 8 oz/about 4) swedes, peeled
    and cut into 4 cm (1½ inch) pieces
125 ml (4 fl oz/½ cup) pouring
    (whipping) cream
2 eggs, lightly beaten
1 egg yolk
3 tablespoons plain (all-purpose) flour
½ teaspoon freshly grated nutmeg
a small pinch of ground cloves
100 g (3½ oz/1¼ cups) fresh breadcrumbs
50 g (1¾ oz) unsalted butter, chopped
4 sage leaves, finely chopped

Preheat the oven to 180°C (350°F/Gas 4). Grease a shallow 18 x 30 cm (7 x 12 inch) baking dish.

Cook the swedes in boiling salted water for 40 minutes, or steam them for 25 minutes, until tender. Drain well and return to the saucepan, then mash well using a potato masher — the mixture should still have some texture.

Place the saucepan back over medium heat and cook the swede, stirring often, for 5–7 minutes, or until the excess liquid has evaporated. Remove from the heat and allow to cool slightly.

Stir in the cream, eggs, egg yolk, flour and spices, then season with sea salt and freshly ground black pepper. Pour the mixture into the baking dish, smoothing the top even.

Put the breadcrumbs and butter in a food processor and blend until fine clumps form, then add the sage and sprinkle the mixture over the swede. Bake for 30–35 minutes, or until golden and set in the middle. Serve hot or warm as a side dish to accompany roasted or grilled meats.

VARIATION: You can substitute the same weight of turnips for swedes in this recipe, if you prefer.

PARSNIP AND PECAN FRITTERS
SERVES 4 AS A STARTER OR SIDE DISH

CHIVE AND SWEET CHILLI CREAM
200 g (7 oz/heaped ¾ cup) sour cream
1 tablespoon finely snipped chives
1 teaspoon lemon juice
2 tablespoons sweet chilli sauce
3–4 drops of Tabasco sauce

3 parsnips (375 g/13 oz in total)
1 egg
3 tablespoons plain (all-purpose) flour
1 tablespoon chopped parsley
50 g (1¾ oz) unsalted butter, melted
3 tablespoons milk
65 g (2¼ oz/⅔ cup) pecans, coarsely
   chopped
a pinch of cayenne pepper
250 ml (9 fl oz/1 cup) vegetable oil

In a small bowl, mix together all the ingredients for the chive and sweet chilli cream. Cover and refrigerate until required.

Bring a saucepan of salted water to the boil. Peel the parsnips, cut them into chunks and immediately add them to the boiling water. Reduce the heat and simmer for 15–20 minutes, or until tender, then drain and allow to cool slightly.

Meanwhile, preheat the oven to 120°C (235°F/Gas ½).

Purée the parsnip using a potato ricer or a food mill, discarding the fibres. Transfer to a bowl and mix in the egg, flour, parsley, butter, milk, pecans and cayenne pepper. Season with sea salt and freshly ground black pepper.

Heat the oil in a heavy-based frying pan over medium heat. Working in batches, add heaped tablespoons of the fritter mixture, flattening them slightly using the back of a spoon. Cook for 15–20 seconds on each side, or until golden. Remove with a slotted spoon and drain on paper towels, then place in the oven to keep warm while cooking the remaining fritters. Serve hot, with the chive and sweet chilli cream.

PARSNIP GNOCCHI
SERVES 4

4 parsnips (500 g/1 lb 2 oz in total)
185 g (6½ oz/1½ cups) plain
  (all-purpose) flour
50 g (1¾ oz/½ cup) grated parmesan
  cheese

GARLIC HERB BUTTER
100 g (3½ oz) unsalted butter
2 garlic cloves, crushed
3 tablespoons chopped lemon thyme
1 tablespoon finely grated lime zest

Bring a saucepan of salted water to the boil. Peel the parsnips, cut them into large chunks and immediately add them to the boiling water. Reduce the heat and simmer for 30 minutes, or until very tender. Drain thoroughly and leave to cool slightly.

Mash the parsnip in a bowl until smooth, or push through a food mill. Sift the flour into the bowl and add half the parmesan. Season to taste with sea salt and freshly ground black pepper and mix to form a soft dough.

Divide the dough in half. Using floured hands, roll out each half on a lightly floured surface into a sausage shape 2 cm (¾ inch) wide. Cut each sausage into short pieces, shape each piece into an oval, then press the top of each gnocchi gently with the prongs of a floured fork.

Bring a large saucepan of salted water to the boil. Add the gnocchi in batches, being careful not to crowd the pot. Stir gently and cook for 1–2 minutes, or until the gnocchi rise to the surface. Remove with a slotted spoon and transfer to warmed shallow serving bowls.

Meanwhile, combine all the garlic herb butter ingredients in a small saucepan and cook over medium heat for 3 minutes, or until the butter is brown and smells nutty. Season to taste with sea salt.

Drizzle the garlic herb butter over the gnocchi, sprinkle with the remaining parmesan and serve.

Celeriac remoulade

SERVES 4–6 AS A STARTER OR SIDE DISH

MUSTARD MAYONNAISE
2 egg yolks
1 tablespoon white wine vinegar or
    lemon juice
1 tablespoon dijon mustard
150 ml (5 fl oz) olive oil

juice of 1 lemon
1 large or 2 small celeriac
2 tablespoons capers, rinsed and drained
5 gherkins (pickles), chopped
2 tablespoons finely chopped parsley
crusty bread, to serve

To make the mustard mayonnaise, whisk the egg yolks, vinegar and mustard together in a bowl. Whisking constantly, add the olive oil, 1 teaspoon at a time, until the mixture begins to thicken, then add the remaining oil in a thin, steady stream. Set aside until required, placing plastic wrap directly on the surface of the mayonnaise to prevent a skin forming.

Pour 1 litre (35 fl oz/4 cups) cold water into a large bowl and add half the lemon juice. Trim and peel the celeriac, then coarsely grate and place in the acidulated water to prevent discolouration. Bring a saucepan of water to the boil over high heat and add the remaining lemon juice. Drain the celeriac and add to the saucepan. After 1 minute, drain the celeriac again and cool under running water. Pat dry with paper towels.

Toss the celeriac with the mustard mayonnaise, capers, gherkins and parsley. Serve with crusty bread.

Celeriac and potato mash

SERVES 4 AS A SIDE DISH

1 tablespoon lemon juice
1 celeriac
1 large roasting potato, such as russet
    (idaho) or king edward, peeled and cut
    into 2.5 cm (1 inch) chunks
250 ml (9 fl oz/1 cup) milk
20 g (³/₄ oz) unsalted butter, softened

Pour 500 ml (17 fl oz/2 cups) cold water into a large bowl and add the lemon juice. Trim and peel the celeriac, then chop into 2.5 cm (1 inch) chunks, placing them in the acidulated water as you go to prevent discolouration.

Drain well, then place in a saucepan with the potato and milk and bring to the boil over high heat. Cover and cook for 15 minutes, or until the celeriac and potato are tender. Mash well, season to taste with sea salt and freshly ground black pepper, stir in the butter and serve immediately.

VARIATION: To make celeriac and apple mash, add two peeled, cored and coarsely chopped apples to the potato and celeriac; boil with the milk and continue as above.

CARROT, SPICE AND SOUR CREAM CAKE
SERVES 8–10

310 g (11 oz/2$^{1}/_{2}$ cups) self-raising flour
2 teaspoons ground cinnamon
1 teaspoon ground nutmeg
150 g (5$^{1}/_{2}$ oz/$^{3}/_{4}$ cup) dark brown sugar
200 g (7 oz/1$^{1}/_{3}$ cups) grated carrot
4 eggs
250 g (9 oz/1 cup) sour cream
250 ml (9 fl oz/1 cup) vegetable oil

ORANGE CREAM CHEESE ICING
3 tablespoons cream cheese, softened
20 g ($^{3}/_{4}$ oz) unsalted butter, softened
1 teaspoon grated orange zest
2 teaspoons orange juice
125 g (4$^{1}/_{2}$ oz/1 cup) icing (confectioners')
  sugar

Preheat the oven to 160°C (315°F/Gas 2–3). Grease a deep, 22 cm (8$^{1}/_{2}$ inch) round cake tin and line the base with baking paper.

Sift the flour, cinnamon and nutmeg into a large bowl, then stir in the sugar and carrot until well combined.

In a bowl, beat together the eggs, sour cream and oil. Add to the flour mixture and stir until well combined. Spoon the batter into the cake tin and smooth the surface even.

Bake for 1 hour, or until a cake tester inserted into the centre of the cake comes out clean. Remove from the oven and allow to cool in the tin for 10 minutes, before turning out onto a wire rack to cool completely.

To make the orange cream cheese icing, beat the cream cheese, butter and orange zest and juice in a bowl using electric beaters until light and fluffy. Gradually add the icing sugar and beat until smooth. Spread the icing over the top of the cooled cake. Cut into slices to serve.

Without the icing, carrot, spice and sour cream cake will keep for 4 days, stored in a cool place in an airtight container, or can be frozen in an airtight container for up to 3 months. The iced cake will keep for 2 days, stored in a cool place in an airtight container.

crisper

From the moment of harvesting, all produce starts to deteriorate. Cooling, however, slows this process. The crisper is ideal for storing the produce in this chapter as it is more humid than the rest of the fridge. Keep your crisper very clean as bacteria and mould spores can quickly accumulate — and don't stuff it too full or air won't be able to circulate.

BEANS

'Green' bean varieties may in fact be yellow, purple or russet, but they are all called 'green' beans because they are picked when young and tender. Yellow (or 'wax') beans have a more subtle flavour than green-skinned beans, while purple ones turn green when cooked. Most modern hybrids are stringless.

A truly fresh bean will break with a clean 'snap' when bent; the pointed end should be straight and not withered. Pods should be smooth and unblemished, and the seeds within just slight bumps; if they are swollen they are overmature and will be mealy rather than sweet. Beans will keep in the crisper, in a ventilated plastic bag, for up to 3 days.

To prepare beans, remove the string, if there is one, and trim the stem end. Depending on their size, they can be cooked whole or cut into lengths. They are usually boiled or steamed, although in Middle Eastern recipes they are often braised in a sauce until very soft. Tender green beans can be roasted whole in some olive oil, intensifying their sweetness, and are delicious in salads, as an antipasto, or as a side dish for meats.

Beans pair well with butter, olive oil, garlic, dill, marjoram, thyme, tomato, potato, peas and other greens, olives, tuna, walnuts, almonds, crisp breadcrumbs, parmesan cheese, anchovies and curry spices.

BEETROOT (BEETS)

Once, only the leaves and stems of the beetroot were eaten, with the large, bulbous roots used strictly medicinally. Now, quite wastefully, we tend to use just the roots, even though the tops are delicious steamed, blanched or braised, dressed with balsamic vinegar and good olive oil.

Beetroot hybrids come in various colours (orange, pink, or with pink and white rings through the flesh) and forms, from very tiny to very elongated or bulbous. The standard burgundy-coloured beets have the strongest flavour and owe their colour to a pigment called betacyanin, to which some people are intolerant, and which stains fingers, clothes and work surfaces.

Available year round, beetroot are usually sold in bunches with the leaves still attached; bright, perky leaves indicate freshness. Choose bulbs of a similar size to make cooking easier — they should have smooth, unblemished skin. Use leaves and stems within 2 days, although trimmed bulbs will keep for 7–10 days in a loosely tied plastic bag in the crisper.

Beetroot is usually boiled with the skin on, to stop the juices 'bleeding'. Leave the long root untrimmed, and a little stem attached. Depending on size, they can take an hour or more to gently boil to tenderness; baby beetroot may only take about 30 minutes. Beetroot also roast beautifully, concentrating the sweet, earthy flavours — trim, wash and dry the beetroot, brush the skins with olive oil and bake in a moderate oven for 1½ hours, or until tender. Leave to cool slightly; the skins (as with boiled beets) should just slip off.

Beetroot is wonderful in a savoury marmalade or relish with red meats. It is especially good with chives, dill, parsley, coriander (cilantro), cloves, cinnamon, red wine, vinegar, orange, sour cream, mustard, horseradish, ginger, walnuts, apples, radicchio, rocket (arugula) and mayonnaise.

BROCCOLI

A member of the brassica family, broccoli is particularly rich in minerals and vitamins A and C. There are several types and hybrids, including a dark blue-green Italian variety, but the broccoli most of us cook and love is the deep, vivid-green sort that peaks in availability from autumn through winter.

Look for very fresh, tightly packed heads, with unopened flower buds and no yellow blooms in evidence. Any leaves should be deep green and undamaged. Store broccoli in a ventilated plastic bag in the crisper for up to 3 days, but don't wash it before storing as water will cause it to spoil.

The stalks can be eaten, but first trim off the thickest part. They require longer cooking than the florets and may need peeling. When trimming the head into florets, leave some stalk attached to hold them together. Briefly steaming the florets until just tender (3–4 minutes is usually sufficient) is preferable to boiling, as boiling makes the florets squelchy.

Broccoli is lovely with almonds, hazelnuts, pine nuts, garlic, ginger, cheese (blue, goat's, cheddar, gruyère, parmesan, ricotta), sour cream, anchovies, tomato, olives, potato, capers, lemon, orange, bacon and ham.

BRUSSELS SPROUTS

Many people boil this much-maligned vegetable until it is mushy, drab and odorous — but when properly prepared, Brussels sprouts have an alluring nutty sweetness and mustardy tang. Like other members of the brassica family, they contain compounds that produce hydrogen sulphide when heated. Brussels sprouts are particularly high in these compounds, which increase in intensity the longer they are cooked, giving an unpleasant smell and flavour.

Brussels sprouts are like a cabbage in miniature, except they grow as compact groups of buds along a stem. They are in season during winter, persisting into spring. Choose the smallest ones you can find as they'll be sweeter and require the least cooking. They should be very firm and a lovely bright green, with no mottled or yellowing outer leaves. Keep them in the crisper in a loosely sealed or perforated plastic bag for no more than 3 days; their flavour gets stronger the longer they are stored.

Steaming is a better method of cooking than boiling so they don't become waterlogged; this may take 7–10 minutes, depending on their size, or less if they are halved. They're also delicious sautéed or cooked in a creamy sauce; lightly steam them first so they're tender in the middle.

Brussels sprouts pair particularly well with chestnuts, bacon, prosciutto, pancetta, roast pork, mustard, horseradish, cream, sour cream, caraway seeds, chives, parsley, garlic, nutmeg and shallots.

CABBAGE

A highly prolific crop, and another member of the brassica family, cabbage is most delicious in the hands of a sensitive cook. There are over 400 types, ranging from round to conical and from green or white to red or purple. The most common is probably the pale, round **green cabbage**. The **savoy** has darker green, crinkled outer leaves and a paler, tender heart, with a mild flavour. It is excellent in cooked dishes, but not so good raw. **Red cabbage** tastes similar to green cabbage and has the same crisp texture and glossy, smooth leaves; they are both great for cooking, pickling and for using raw.

The pigment in red cabbage is water soluble and leeches out when cooked; it also reacts to alkali (such as hard tap water) and turns blueish purple, which is why it often discolours when cooked. To preserve its lovely hue, cook red cabbage with a splash of something acidic, such as lemon juice or vinegar.

All cabbages are rich in vitamin C, various minerals and cancer-inhibiting compounds. They contain mustard-like chemicals which are released when the cells are broken (by cutting or chewing), hence their somewhat 'hot' taste. Prolonged cooking causes sulphur substances to be produced, as anyone who has smelt overcooked cabbage will know!

When buying green or red cabbage, look for tight, heavy heads. Make sure there are no brown or yellow patches and that the outer leaves tightly enclose the head. Savoy cabbages will have slightly looser leaves; the outer leaves should be a darkish green, with no wilting or brown spots. Savoy cabbage only keeps for up to 3 days in a loosely sealed plastic bag in the crisper; red and green cabbages will keep for 1 week, perhaps longer.

To use cabbage, discard any tough outer leaves and remove the central 'core'. If you need the leaves whole for stuffing, carefully remove each leaf from the base — savoy leaves are best for this as they are looser and more easily removed. Wash cabbage well as the leaves can harbour caterpillars.

Cabbage can be steamed, gently boiled, sautéed, braised, stir-fried, pickled, or served raw in salads. Cabbage pairs beautifully with pork, bacon, sausages, mustard, butter, cream, sour cream, dill, sage, parsley, apple, vinegar, raisins, onions, mayonnaise and gherkins (pickles).

Cauliflower

Cauliflower and broccoli are from the same family, yet the two are quite distinct. The tight, knobbly part of a cauliflower head is called the 'curd', and comprises immature flower buds, surrounded by a tight network of green leaves which shield the curd from the sun. Like other plants in the brassica family, cauliflower contains vitamin C, potassium, folate and antioxidants. It also contains purines, which should be avoided by those with gout.

Winter is peak cauliflower season, although it is generally available year round. The white-headed cauliflower is the most common, but you may see purple or green-headed varieties as well. Look for cauliflower with very tight, unblemished curds and tight leaves (ones with leaves will also keep longer). Store in the crisper stem side down to prevent condensation, in a loosely sealed or perforated plastic bag. It will stay fresh for 4–5 days.

To prepare, cut off the tough stalks and leaves. Cut off florets from the thick base; the larger they are, the longer they'll take to cook. Keep the stems and tender leaves — when chopped and sautéed, they're a nutritious addition to soups, risottos, vegetable braises and purées. Steaming until just tender (8–12 minutes) is better than boiling, or try roasting the florets in some butter or olive oil in a moderate oven for about 30 minutes; they'll turn deep brown and take on a nutty flavour perfect for purées or salads. Avoid cooking cauliflower in iron or cast-iron pans as these can discolour it.

Use cauliflower in soups (especially puréed creamy ones, where it pairs well with potato), salads, pasta dishes and risottos. It is delicious with golden, buttered breadcrumbs, olives, caramelised onions, anchovies, cream, chives, dill, cumin seeds, mustard, parsley, capers, cheese (parmesan, cheddar, gruyère, blue and smoked) and bacon.

CELERY

Celery was originally used for medicinal purposes, and modern science has indeed confirmed that it contains compounds called pthalides, which help reduce blood pressure. Of course it is also incredibly delicious!

Celery is available year round. The sturdy outer stalks are fibrous and bright green; the very interior stalks are collectively known as the 'heart' and are prized in cooking and salads. The tender inner leaves are also good to eat (unlike the larger, outer green ones, which can be used in stocks). You can use a potato peeler to remove very stringy fibres from large stalks.

Try to buy celery with the leaves intact as these give the best indication of freshness. Outer stalks should be shiny, erect and snap crisply if bent. Celery will keep, stored in the crisper in a plastic bag, for up to 1 week.

Celery is most often used as a salad vegetable or as a background vegetable in soups, stews, sauces, stocks and braises. It can, however, star in its own right when braised in stock, plenty of olive oil and some aromatics.

Celery is fabulous with apple (as in a waldorf salad; see page 88), lemon, pears, cream, cheese (blue, ricotta, cream cheese, gruyère), mayonnaise, vinegar, tomato, anchovies, raisins, almonds, walnuts, pine nuts, chives, parsley, thyme, white wine, tinned tuna, fish, corned beef, chicken and lamb.

CUCUMBER

Cucumbers are most abundant in summer, when their fresh, crisp, fragrant coolness shines in salads, chilled soups, salsas and pickles. If you can find them, 'heirloom' varieties have far more character and flavour than the mass-produced supermarket varieties. A 'proper' cucumber should taste deeply herbal and a little sweet; organically grown ones are often your best bet.

There are many varieties, ranging from the tiny, deep-green **gherkins (pickles)** used for pickling, to large round 'apple' cucumbers, with their pale, creamy-coloured skin. Some cylindrical types can grow up to 60 cm (2 feet) long. The most common varieties are the **telegraph (long) cucumber**, which is often sold wrapped in plastic and has thin skin and few seeds; the **garden** or **green ridge cucumber**, which has larger seeds and thicker skin that needs peeling; and the small **Lebanese (short) cucumber**, which has tender skin, sweet pale green flesh and scant, soft seeds.

Cucumbers are around 96% water, so they should feel quite heavy for their size and also firm. Those that are small and slender for their type are usually sweeter than large specimens. Cucumbers will keep for 4–5 days in the crisper, but for the best flavour bring to room temperature before using.

Some cucumbers have very bitter skin, but not all require peeling. Some cucumbers have seeds that are too large to comfortably eat; simply cut them in half lengthways and scoop out the seeds using a teaspoon. Try not to chop, slice or mix cucumber with other ingredients until just before using as it will release juice upon standing.

Cucumber complements seafood and marries well with cream, sour cream, crème fraîche, feta and goat's cheese, olive, tomatoes, vinegar, lemon, orange, watercress, peas, spring onions (scallions), sesame, cumin, coriander (cilantro), dill, tarragon, oregano, parsley and chives.

Fennel

Beyond its traditional Mediterranean stronghold, fennel is becoming increasingly popular for its subtle aniseed flavour. Fennel appears in both autumn and spring, ranging in size from diminutive 'baby' bulbs to rather hefty family-size ones. Its fleshy leaf stems can be eaten cooked or raw.

The green fronds are an excellent indicator of freshness; they should be brightly coloured, with no sign of wilting. The bulbs should be firm and creamy white, with the layers tightly packed together. They will keep in a perforated or loosely sealed plastic bag in the crisper for 3–4 days.

Smaller bulbs tend to be milder in flavour and can be eaten raw as a crudité or in salads; slice them paper-thin at the last moment to avoid browning. Larger bulbs are more suited to cooking, either by braising, roasting, grilling (broiling), barbecuing or being grated into fritters. Fennel is also good in pasta dishes, risotto, mixed vegetable stews and soups.

Discard any hard and fibrous outer white leaves, then cut the bulb in half lengthways and cut away the hard core, before cutting into wedges, quarters or thin slices. To stop sliced fennel browning, slip it into water spiked with lemon juice. The green fronds from baby fennel can be chopped finely and used in salads or risottos, but discard those from larger bulbs.

Fennel pairs beautifully with cream, chicken stock, olive oil, lemon, orange, cheese (blue, goat's and parmesan), almonds, pine nuts, hazelnuts and walnuts, tomato, potato, anchovy, tuna, mayonnaise, capers, olives, radicchio, watercress, smoked salmon and prosciutto.

LEEK

Leeks were highly esteemed in times past, and many food cultures still adore them. In common with other members of the onion family, the humble leek has health-giving properties, helping guard against high blood pressure and certain cancers. Their true season is in late spring through to autumn.

Thinner leeks tend to be the sweetest and least fibrous. Choose very firm, heavy, straight leeks with dark green, healthy leaves. Fresh leeks will have bright white roots with dirt still attached. Store in the crisper (with the leaves still attached) in a loosely sealed plastic bag for up to 1 week.

The white part of the leek is the most tender. For cooking, cut off the green tops (these can be washed well, then chopped and used in stocks), down to the white stem. To serve leeks whole, keep the root end intact and make deep cuts down through the top 7 cm ($2^3/4$ inches) or so into the white, then wash well, pulling back the layers to get to where the dirt is trapped; otherwise, simply chop to the desired size and rinse well in cold water. Leeks can be steamed, braised, sautéed and stir-fried, or stewed in their own juices in a covered pan with a little butter, nutmeg and herbs.

Leeks are lovely with chicken and chicken stock, fish and fish stock, cream, cheese (cheddar, gruyère, blue and parmesan), dill, chives, parsley, thyme, marjoram, black pepper, potatoes, parsnips, anchovies, mustard, walnuts and hazelnuts, and the oils from those nuts.

MUSHROOMS

The mushroom is the fruit of a fungus, with varieties numbering in the thousands. Although their season is autumn to early winter, some are grown commercially year round. The white mushroom we are most familiar with (*Agaricus bisporus*) was first cultivated in Paris in the 1800s. Brown-capped versions of it are variously called Swiss browns, crimini or portobello, depending on their size and which country you are in. Other cultivated mushrooms include shiitake, oyster, enoki and hen-of-the-woods. Porcini (also called ceps) and morels, prized for their deep, earthy flavours, are foraged from the wild.

Mushrooms dry out very quickly. Easily damaged by condensation, they are best stored in a paper (rather than plastic) bag. Look for firm mushrooms with dry-ish (but not dried-out) caps. Gills should look fresh and healthy, not shrivelled, slimy or dried. In a paper bag in the crisper they'll keep up to 3 days.

Avoid washing mushrooms — any dirt can simply be wiped off, or if they are very grubby, peel the skin off using your fingers, working from the edge of the mushroom towards the centre. Generally, the flatter a mushroom, the older it is; tight 'button' mushrooms are the youngest. With their stronger flavour, older mushrooms are nicer cooked than raw.

Mushrooms are delicious with beef, chicken, liver, bacon, pine nuts, hazelnuts, parmesan and goat's cheese, sage, rosemary, oregano, basil, garlic, currants, anchovies, cream, sour cream, brandy, sherry and puff pastry.

Silverbeet (swiss chard)

Native to the Mediterranean, silverbeet is related to the beetroot, hence the name 'beet'. Although it has high oxalic acid levels, it is also very rich in iron, magnesium, potassium and vitamins A, K and C, making it one of the most nutritious vegetables. It is in season from spring through autumn.

Buy bunches with sturdy clean, bright stems and upright, undamaged leaves. The leaves should be a deep, glossy green, with thick, bright-silvery ribs and veins. Some varieties have red, pink or orange ribs; their juice will stain everything it comes in contact with. Silverbeet will keep in the crisper, loosely tied in a plastic bag, for 2–3 days.

If using the stems, one bunch is enough for 3–4 people, but if using just the leaves, you may need an extra bunch. The leaves are usually just steamed, but you can also cook them, in a tightly closed saucepan, in just the water that clings to the leaves after washing. The stems discolour when cut, so put them in a bowl of acidulated water as you are preparing them (some cooks also remove any thick fibres using a small sharp knife). The stems take about 15 minutes to cook, the leaves just 3–4 minutes. Blanched leaves can be rolled around fillings and baked.

Silverbeet pairs well with lentils, chickpeas, bacon, garlic, cumin, tomatoes, ginger, cream, blue cheese, walnuts, hazelnuts, lemon, saffron, red wine vinegar, chilli flakes, anchovies, pine nuts, raisins and olive oil.

Spinach

This tender vegetable stars in everything from soups, frittatas, soufflés and pies to pasta dough and quiche florentine. Spinach is available all year, but its true season is winter through to early spring.

Choose bunches with tender leaves that are upright, unblemished, broad in shape and a lovely deep jade. Store in a sealed plastic bag in the crisper for 2 days; if you have too much, trim, blanch and freeze for up to 3 months to use later in pies, pasta and rice dishes.

To use, discard the tough stems, then gently submerge the leaves in cool water and drain well. (If you need to dry them, gently pat with paper towels.) If you are steaming spinach, the water clinging to the leaves is all the liquid you'll need; tightly cover the pan and shake it occasionally, and don't cook for longer than about 3 minutes or it will become slimy.

Spinach is great with eggs, chicken, fish, ham, liver, mushrooms, potatoes, garlic, cream, sour cream, yoghurt, cheese, olive oil, good vinegar, soy sauce, ginger, sesame seeds, nuts, rice, pastry, citrus, nutmeg and paprika.

Rhubarb

Rhubarb is a spring vegetable, but most of us cook it as a fruit. Only the striking red stalks are edible; the large attractive leaves contain large quantities of oxalic acid, which can cause kidney stones. Stalks vary in hue from quite green to deep ruby red; generally, redder stalks are sweetest.

Choose stalks that are firm, smooth and upright, with leaves attached if possible, but avoid very thick stalks as these can be fibrous. Sealed in plastic, they will keep in the crisper for 1 week.

To prepare, trim the stalks at the base, rinse them, then cut to the required length. The juice stains terribly, so take care not to get it on your clothing. Some cooks like to strip the outer fibre from the uncut stems.

Rhubarb goes particularly well with apples, plums, strawberries and raspberries, orange juice and zest, vanilla, ginger, cinnamon, honey and flower waters (rose and orange). Simply stewed with sugar and served hot or chilled, rhubarb is heavenly with whipped cream, vanilla ice cream, creamy rice pudding, custard, mascarpone or lightly sweetened ricotta cheese.

Zucchini (courgette)

A member of the squash family, zucchini is a summer vegetable with deep green, pale green or bright yellow skin. There are varieties of summer squash which are round and flat and very similar to zucchini in flavour and uses.

Zucchini grow on a vine, and if left to develop become very large, with bland, watery flesh. The smaller, tender specimens, on the other hand, have sweeter, firmer flesh and soft, edible skin and seeds. Choose zucchini with smooth, slightly shiny skin. The interior should be creamy and bright. Zucchini lose flavour quickly, so store them for no longer than 2 days in the crisper, in a perforated or loosely tied plastic bag.

To prepare, simply rinse, trim the ends and slice to the desired size, or grate or cut into thin ribbons. Zucchini are best cooked quickly, either by steaming, sautéeing, stir-frying or grilling (broiling), although whole zucchini can be boiled. Zucchini are also good raw, either thinly sliced or grated.

They are superb with butter, olive oil, garlic, onion, lemon, cheese (feta, parmesan, pecorino), anchovies, mild vinegar, Mediterranean vegetables, pine nuts, almonds, currants, basil, mint, oregano, thyme and parsley.

BROCCOLI AND RICOTTA SOUFFLE
SERVES 4

60 g (2¼ oz/1 cup) small broccoli florets
2 tablespoons olive oil
40 g (1½ oz) unsalted butter
1 onion, finely chopped
1 garlic clove, crushed
400 g (14 oz/scant 1⅔ cups) ricotta
  cheese
50 g (1¾ oz/½ cup) grated parmesan
  cheese
5 egg yolks, lightly beaten
a pinch of nutmeg
a pinch of cayenne pepper
5 egg whites
a pinch of cream of tartar
3 tablespoons dry breadcrumbs

Preheat the oven to 190°C (375°F/Gas 5).

Cook the broccoli florets in boiling salted water for 4 minutes, then drain well and roughly chop.

Heat the olive oil and butter in a frying pan. Add the onion and garlic and sauté over medium heat for 5 minutes, or until the onion has softened. Transfer to a large bowl and add the broccoli, ricotta, parmesan, egg yolks, nutmeg and cayenne pepper. Season with sea salt and freshly ground black pepper. Mix well.

In a clean, dry bowl, whisk the egg whites with the cream of tartar and a pinch of sea salt until stiff peaks form. Stir one-third of the beaten egg white into the broccoli mixture to loosen, then gently fold in the remaining egg white.

Grease a 1 litre (35 fl oz/4 cup) soufflé dish. Sprinkle with the breadcrumbs, turn the dish to coat, then shake out the excess. Spoon the broccoli mixture into the dish and bake for 35–40 minutes, or until puffed and golden brown. Serve immediately.

NOTE: Because this soufflé is based on ricotta cheese it won't rise as much as a conventional soufflé.

CABBAGE ROLLS
MAKES 12 LARGE ROLLS

270 ml (9 ½ fl oz) olive oil
1 onion, finely chopped
a large pinch of allspice
a large pinch of ground nutmeg
1 teaspoon ground cumin
2 bay leaves
1 large head of cabbage
500 g (1 lb 2 oz) minced (ground) lamb
220 g (7¾ oz/1 cup) short-grain white
   rice
4 garlic cloves, crushed
4 tablespoons toasted pine nuts
2 tablespoons chopped mint
2 tablespoons chopped flat-leaf (Italian)
   parsley
1 tablespoon chopped currants
4 tablespoons lemon juice
1 teaspoon sea salt
extra virgin olive oil, for drizzling
lemon wedges, to serve

Heat 1 tablespoon of the olive oil in a saucepan. Add the onion and sauté over medium heat for 10 minutes, or until golden. Add the allspice, nutmeg and cumin and cook for a further 2 minutes, or until fragrant. Transfer to a large bowl.

Bring a very large saucepan of water to the boil and add the bay leaves. Cut the tough outer leaves and about 5 cm (2 inches) of the inner core from the cabbage, then carefully add the whole cabbage to the boiling water. Cook for 5 minutes, then carefully loosen a whole outer leaf with tongs and remove. Continue to cook and remove the leaves until you reach the core. Drain the remaining cabbage well, reserving the cooking liquid, and leave to cool.

Take 12 leaves of equal size and cut a small 'v' from the core end of each to remove the thickest part, and so each leaf sits as flat as possible. Use three-quarters of the remaining leaves to line the base of a very large heavy-based saucepan, to prevent the rolls sticking to the pan while they cook.

Add the lamb, rice, garlic, pine nuts, mint, parsley and currants to the onion mixture, season with sea salt and freshly ground black pepper and mix together well.

With the core end of a cabbage leaf closest to you, form 2 tablespoons of the lamb mixture into an oval shape and place it in the centre of the leaf. Roll the leaf up, tucking in the sides as you go, forming a neat parcel. Repeat with the remaining 11 leaves and filling. Place the rolls in a single layer in the lined saucepan, seam side down; the rolls should be packed quite tightly.

Mix together 625 ml (21½ fl oz/2½ cups) of the reserved cabbage cooking liquid, the remaining olive oil, lemon juice and sea salt, then pour over the rolls; they should be just covered. Place the remaining cabbage leaves over the top. Cover the saucepan and bring to the boil, then reduce the heat and simmer for 1¼ hours, or until the rice is tender.

Using a slotted spoon, carefully remove the rolls from the pan and arrange on a platter. Drizzle with extra virgin olive oil. Serve hot or at room temperature, with lemon wedges.

BROCCOLI AND PINE NUT SOUP
SERVES 6 AS A STARTER

30 g (1 oz) unsalted butter
1 onion, finely chopped
1.5 litres (52 fl oz/6 cups) chicken stock
750 g (1 lb 10 oz) broccoli, trimmed
4 tablespoons pine nuts, plus extra,
  to serve
toasted focaccia, to serve
extra virgin olive oil, to serve

Melt the butter in a large saucepan. Add the onion and sauté over medium heat for 5 minutes, or until softened but not browned. Add the stock and bring to the boil.

Cut the florets from the broccoli and set aside. Chop the broccoli stalks and add them to the pan, then reduce the heat, cover and simmer for 15 minutes. Add the florets and simmer, uncovered, for 10 minutes, or until the florets are tender. Remove from the heat and allow to cool completely.

Add the pine nuts to the soup. Transfer to a blender or food processor, in batches if necessary, and blend until smooth. Season to taste with sea salt and freshly ground black pepper, then gently reheat.

Serve sprinkled with extra pine nuts, and with some toasted foccacia drizzled with extra virgin olive oil.

CAULIFLOWER FRITTERS
SERVES 4–6 AS A STARTER OR SIDE DISH

55 g (2 oz/1/2 cup) besan (chickpea flour)
  (see Note)
1/2 teaspoon sea salt, plus extra, for
  sprinkling
2 teaspoons ground cumin
1 teaspoon ground coriander
1 teaspoon ground turmeric
a pinch of cayenne pepper, plus extra,
  for sprinkling (optional)
1 egg
1 egg yolk
600 g (1 lb 5 oz) cauliflower, cut into
  bite-sized florets
oil, for deep-frying

Sift the besan, sea salt and spices into a bowl. Make a well in the centre. Lightly whisk the egg and egg yolk with 3 tablespoons water, then pour into the well in the flour mixture, whisking until smooth. Set aside to stand for 30 minutes.

Meanwhile, preheat the oven to 120°C (235°F/Gas 1/2).

Fill a large heavy-based saucepan one-third full of oil and heat to 180°C (350°F), or until a cube of bread dropped into the oil browns in 15 seconds. Dip the cauliflower florets into the batter, allowing the excess to drain off, then fry in batches for 3 minutes, or until puffed and browned. Remove with a slotted spoon and drain on paper towels, then place in the oven to keep warm while cooking the remaining fritters. Serve hot, sprinkled with sea salt and extra cayenne pepper, if desired.

NOTE: Besan is available from health food shops and Indian grocers.

CHESTNUT, PANCETTA AND CABBAGE SOUP
SERVES 4

250 g (9 oz) savoy cabbage, roughly
   chopped
2 tablespoons olive oil
1 large onion, finely chopped
185 g (6½ oz) pancetta, diced
3 garlic cloves, crushed
2 tablespoons chopped rosemary
250 g (9 oz) cooked peeled chestnuts
   (see Note)
150 ml (5 fl oz) red wine
extra virgin olive oil, for drizzling

Bring 1.5 litres (52 fl oz/6 cups) salted water to the boil in a large saucepan. Add the cabbage and cook over high heat for 10 minutes. Drain, reserving the cooking liquid. Allow the cabbage to cool slightly, then chop into bite-sized pieces.

Heat the olive oil in a large saucepan. Add the onion and pancetta and sauté over medium–high heat for 5 minutes, or until the onion is soft and the pancetta lightly browned. Add the garlic and rosemary and cook for a further 2 minutes.

Using your hands, break up the chestnuts and add them to the pan with half the cabbage. Stir together, then season to taste with sea salt and freshly ground black pepper. Add the wine, bring to the boil and cook for 2 minutes. Add the reserved cabbage cooking liquid, reduce the heat and simmer for 15 minutes.

Remove half the mixture from the pan and allow to cool slightly, then transfer to a food processor and blend to a purée.

Mix the purée back into the soup, with the remaining cabbage. Serve with a drizzle of extra virgin olive oil.

NOTE: Cooked chestnuts are available from delicatessens, sold vacuum-packed, frozen or in tins.

CAULIFLOWER PILAFF

SERVES 4–6 AS A SIDE DISH

200 g (7 oz/1 cup) basmati rice
2 tablespoons olive oil
1 large onion, thinly sliced
$\frac{1}{4}$ teaspoon cardamom seeds
$\frac{1}{2}$ teaspoon ground turmeric
1 cinnamon stick
1 teaspoon cumin seeds
$\frac{1}{4}$ teaspoon cayenne pepper
500 ml (17 fl oz/2 cups) vegetable or
   chicken stock
800 g (1 lb 12 oz) cauliflower, trimmed
   and cut into florets
2 large handfuls of coriander (cilantro)
   leaves, chopped

Rinse the rice under cold running water, then drain well in a sieve and set aside.

Heat the olive oil in a saucepan. Add the onion and sauté over medium heat for 5 minutes, or until softened and light golden. Add all the spices and cook, stirring, for 1 minute.

Add the rice to the pan and stir to coat in the spices. Stir in the stock and cauliflower, then cover and bring to the boil. Reduce the heat to very low and gently simmer for 15 minutes, or until the rice and cauliflower are tender and all the stock has been absorbed.

Stir in the coriander and serve.

PICCALILLI

FILLS SIX 250 ML (9 FL OZ/1 CUP) JARS

400 g (14 oz) cauliflower, chopped
1 small cucumber, chopped
200 g (7 oz) green beans, trimmed
  and chopped
1 onion, chopped
2 carrots, chopped
2 celery stalks, chopped
4 tablespoons sea salt
220 g (7³/₄ oz/1 cup) sugar
1 tablespoon mustard powder
2 teaspoons ground turmeric
1 teaspoon ground ginger
1 red chilli, seeded and finely chopped
1 litre (35 fl oz/4 cups) white vinegar
200 g (7 oz) frozen broad (fava) beans,
  peeled
60 g (2¹/₄ oz/¹/₂ cup) plain (all-purpose)
  flour

Put the cauliflower, cucumber, beans, onion, carrot, celery and sea salt in a large bowl. Add enough water to just cover the vegetables, then cover with an upturned small plate to keep the vegetables submerged. Leave to stand overnight.

Drain the vegetables well, rinse under cold running water, then drain again. Place in a large saucepan with the sugar, mustard powder, turmeric, ginger, chilli and all but 185 ml (6 fl oz/³/₄ cup) of the vinegar. Bring to the boil, then reduce the heat and simmer for 3 minutes. Stir in the broad beans.

Blend the flour with the remaining vinegar and stir it into the vegetable mixture. Cook, stirring, until the mixture boils and thickens.

Spoon the piccalilli into hot, sterilised jars (see Note below) and seal. Label and date for storage.

Leave for 1 month for the flavours to develop fully. Store in a cool, dark place for up to 6 months.

Piccalilli is delicious with strong cheddar cheese and meats such as roast beef and ham.

NOTE: Jars must always be sterilised before pickles, preserves or jams are put in them for storage, otherwise bacteria will multiply. To sterilise your jars and lids, rinse them with boiling water and place in a warm oven for 20 minutes, or until completely dry. (Jars with rubber seals are safe to warm in the oven and won't melt.) Never dry your jars with a tea towel (dish towel) — even a clean one may have germs on it and contaminate the jars.

OPEN LASAGNE OF MUSHROOMS, PINE NUTS AND THYME
SERVES 4

200 g (7 oz) fresh lasagne sheets
300 g (10½ oz) assorted mushrooms
 (see Note)
80 g (2¾ oz) unsalted butter
1 tablespoon olive oil
2 rindless bacon slices, cut into 4 x 2 cm
 (1¼ x ¾ inch) pieces
2 garlic cloves, finely sliced
1 tablespoon thyme leaves
1 tablespoon pine nuts, lightly toasted
3 tablespoons thick (double/heavy)
 cream
3 tablespoons extra virgin olive oil
4 tablespoons coarsely shredded pecorino
 cheese

Bring a large saucepan of salted water to the boil. Cut the lasagne sheets into sixteen 8 cm (3¼ inch) squares. Boil half the squares for 4 minutes, or until *al dente*. Using a slotted spoon, transfer to a bowl of cold water, leave for 15–20 seconds, then drain. Lay the pasta squares flat on a dry tea towel (dish towel) and cover with another tea towel; it doesn't matter if the squares have cooked to uneven sizes. Repeat with the remaining pasta squares.

Trim the stalks of the mushrooms, then wipe the caps clean using a damp cloth or paper towel.

Heat the butter and olive oil in a large frying pan. Add the mushrooms and bacon and sauté over high heat for 3–4 minutes, or until browned. Add the garlic and thyme and cook for a further minute. Add the pine nuts, cream and 2 tablespoons of the extra virgin olive oil and stir until combined. Remove from the heat and season to taste with sea salt and freshly ground black pepper.

Preheat the oven grill (broiler) to medium–high. Divide four pasta squares among four warmed shallow flameproof pasta bowls or deep plates. Cover with a heaped tablespoon of the mushroom mixture. Repeat the layering twice, then top with the last four pasta squares — the pasta doesn't have to be in uniform stacks, or the piles neat. Finish with the last of the mushrooms, two or three per stack.

Drizzle with the remaining extra virgin olive oil and scatter with the pecorino. Place the bowls under the grill for 1–2 minutes, or until the cheese has melted. Serve hot or warm.

NOTE: You can use your favourite type of cap mushroom — such as chestnut, Swiss brown, portobello or even shiitake — as long as they are no bigger than 4 cm (1½ inches) across.

POLENTA SQUARES WITH MUSHROOM RAGU

SERVES 4

POLENTA
500 ml (17 fl oz/2 cups) vegetable stock
 or water
150 g (5½ oz/1 cup) polenta
20 g (¾ oz) unsalted butter
75 g (2½ oz/¾ cup) grated parmesan
 cheese, plus 3 tablespoons, extra

5 g (¼ oz) dried porcini mushrooms
200 g (6½ oz) Swiss brown mushrooms
300 g (10 oz) field mushrooms
125 ml (4 fl oz/½ cup) olive oil
1 onion, finely chopped
3 garlic cloves, finely chopped
1 bay leaf
2 teaspoons finely chopped thyme
2 teaspoons finely chopped oregano
2 handfuls of flat-leaf (Italian) parsley,
 finely chopped
1 tablespoon balsamic vinegar

Grease a shallow, 20 cm (8 inch) square glass or ceramic baking dish or cake tin.

To make the polenta, put the stock and a pinch of sea salt in a large saucepan and bring to the boil. Stirring constantly, add the polenta in a thin, steady stream. Reduce the heat and cook, stirring frequently, for 15–20 minutes, or until the polenta is very thick and comes away from the side of the pan. Remove from the heat and stir in the butter and parmesan. Spread the mixture into the baking dish, smoothing the surface even. Allow to cool to room temperature, then refrigerate for 20 minutes.

Meanwhile, soak the porcini mushrooms in 125 ml (4 fl oz/½ cup) boiling water for 10 minutes, or until softened. Drain, reserving 4 tablespoons of the soaking liquid.

Trim the stalks of the Swiss brown and field mushrooms, then wipe the caps clean using a damp cloth or paper towel. Thickly slice the Swiss browns, and coarsely chop the field mushrooms.

Heat 4 tablespoons of the olive oil in a large frying pan. Add the mushrooms, including the porcini, and cook over medium heat for 4–5 minutes, or until softened. Remove from the pan and set aside.

Heat the remaining oil in the pan. Add the onion and cook over medium heat for 5 minutes, or until softened. Stir in the garlic and cook for a further minute.

Strain the reserved mushroom soaking liquid and add to the pan with the bay leaf, thyme and oregano. Season with sea salt and freshly ground black pepper and cook for 2 minutes. Return the mushrooms to the pan, add the parsley and vinegar and cook for 1 minute, or until nearly all the liquid has evaporated. Remove the bay leaf and adjust the seasoning, if necessary.

Meanwhile, heat the oven grill (broiler) to medium. Sprinkle the extra parmesan over the polenta and cook under the grill for 10 minutes, or until the polenta is lightly browned and the cheese has melted. Cut into four 10 cm (4 inch) squares.

Place a polenta square in the centre of each serving plate and top with the mushroom ragù. Sprinkle with freshly ground black pepper and serve.

BAKED MUSHROOMS
SERVES 4 AS A SIDE DISH

250 g (9 oz) button mushrooms
200 g (6$\frac{1}{2}$ oz) oyster mushrooms
200 g (6$\frac{1}{2}$ oz) fresh shiitake mushrooms
100 g (3$\frac{1}{3}$ oz) Swiss brown mushrooms
2 tablespoons extra virgin olive oil

TOPPING
80 g (2$\frac{3}{4}$ oz/1 cup) fresh breadcrumbs
    (see Note)
3 tablespoons grated parmesan cheese
2 tablespoons chopped flat-leaf (Italian)
    parsley
1 tablespoon chopped thyme
2 garlic cloves, crushed
1 teaspoon cracked black pepper

Preheat the oven to 180°C (350°F/Gas 4).

Trim the stalks of the mushrooms, then wipe off any dirt using a damp cloth or paper towel. Cut any large mushrooms in half lengthways.

Sprinkle the base of a large baking dish with a little water. Place the mushrooms in a single layer in the dish, stems facing upwards.

Put all the topping ingredients in a bowl and mix together well. Sprinkle the topping over the mushrooms and drizzle with the olive oil. Bake for 12–15 minutes, or until the mushrooms have heated through. Serve warm.

NOTE: Slightly stale day-old bread is perfect for breadcrumbs. Simply remove the crusts and chop the bread in a food processor until crumbs form. Avoid using pre-sliced packaged bread for making breadcrumbs as this type of bread tends not to go stale. For best results use baguette, sourdough or a similar bread procured from a good bakery.

VEGETABLE TIAN
SERVES 6–8 AS A SIDE DISH

1 kg (2 lb 4 oz) red capsicums (peppers)
125 ml (4 fl oz/½ cup) olive oil
2 tablespoons pine nuts
800 g (1 lb 10 oz) silverbeet (Swiss chard), stems removed and the leaves coarsely shredded
freshly ground nutmeg, to taste
1 onion, chopped
2 garlic cloves
2 teaspoons chopped thyme
750 g (1 lb 10 oz) tomatoes, peeled, seeded and diced (see Note)
1 large eggplant (aubergine), cut into 1 cm (½ inch) rounds
5 small zucchini (courgettes), about 500 g (1 lb 2 oz) in total, thinly sliced on the diagonal
3 ripe tomatoes, cut into 1 cm (½ inch) slices
1 tablespoon fresh breadcrumbs
4 tablespoons grated parmesan cheese
30 g (1 oz) unsalted butter, chopped

Preheat the oven grill (broiler) to high.

Cut the capsicums into quarters and remove the seeds and membranes. Grill the capsicums, skin side up, until the skin blackens and blisters. Transfer to a bowl, cover with plastic wrap and leave until cool enough to handle. Slip the blackened skin off the capsicums, then cut the flesh into large strips. Place in a lightly greased 25 x 20 x 5 cm (10 x 8 x 2 inch) baking dish and season lightly with sea salt and freshly ground black pepper.

Preheat the oven to 200°C (400°F/Gas 6). Heat 2 tablespoons of the olive oil in a frying pan. Add the pine nuts and fry over medium heat for 1–2 minutes, or until golden, shaking the pan often so they don't burn. Remove with a slotted spoon and set aside.

Add the silverbeet to the pan and cook for 5 minutes, or until softened. Add the pine nuts and season to taste with sea salt, freshly ground black pepper and nutmeg. Spread the silverbeet mixture over the capsicum slices.

Wipe the pan clean and heat another tablespoon of the olive oil. Add the onion and cook over medium heat for 7–8 minutes, or until soft and golden. Add the garlic and thyme, cook for 1 minute, then add the diced tomato. Bring to the boil, reduce the heat and simmer for 10 minutes. Spread the sauce evenly over the silverbeet.

Wipe the pan clean again and heat the remaining oil. Add the eggplant and cook in batches over high heat for 4–5 minutes on each side, or until golden. Drain on paper towels and place in a single layer over the tomato sauce. Season lightly.

Arrange the zucchini and tomato slices in alternating layers over the eggplant. Sprinkle the breadcrumbs and parmesan over the top, then dot with the butter.

Bake for 25–30 minutes, or until the topping is golden. Serve warm or at room temperature.

NOTE: To peel tomatoes, score a small cross in the base, place in a saucepan of boiling water for 20 seconds, then remove using a slotted spoon and plunge into a bowl of iced water. Drain the tomatoes and peel the skins away from the cross. To remove the seeds, cut the tomatoes in half and scoop out the seeds with a teaspoon.

MUSHROOM PATE ON MINI TOASTS

SERVES 4–6 AS A STARTER

60 g (2¼ oz) unsalted butter
1 small onion, chopped
3 garlic cloves, crushed
375 g (13 oz) button mushrooms,
  quartered
125 g (4½ oz/1 cup) slivered almonds,
  toasted
2 tablespoons pouring (whipping) cream
2 tablespoons finely chopped thyme
3 tablespoons finely chopped flat-leaf
  (Italian) parsley
6 slices of wholemeal (whole-wheat)
  bread

Heat the butter in a large frying pan. Add the onion and sauté over medium heat for 5 minutes, or until softened. Add the garlic and cook for a further minute.

Increase the heat to high, add the mushrooms and cook, stirring frequently, for 5–10 minutes, or until the mushrooms are soft and most of the liquid has evaporated. Remove from the heat and allow to cool.

Using a food processor, roughly chop the almonds, then add the mushroom mixture and process until smooth. With the motor running, pour in the cream. Stir in the herbs, then season to taste with sea salt and freshly ground black pepper.

Spoon the mixture into two 250 ml (9 fl oz/1 cup) ramekins and smooth the surface. Cover and refrigerate for 4–5 hours to allow the flavours to develop.

Preheat the oven to 180°C (350°F/Gas 4). Remove the crusts from each slice of bread. Cut each slice in half on the diagonal, then cut in half again, to form triangles. Place in a single layer on a large baking tray, then bake for 20–25 minutes, or until crisp.

Serve immediately, with the mushroom pâté.

ROAST BEEF AND SPINACH SALAD WITH HORSERADISH CREAM
SERVES 4

HORSERADISH CREAM
125 g (4 oz/½ cup) Greek-style yoghurt
1 tablespoon bottled horseradish
2 tablespoons lemon juice
2 tablespoons pouring (whipping) cream
2 garlic cloves, crushed
2–3 dashes of Tabasco sauce

200 g (7 oz) green beans, trimmed
500 g (1 lb 2 oz) rump steak, cut into
    slices 3 cm (1¼ inch) thick
1 red onion, cut in half
1 tablespoon olive oil
100 g (3½ oz/2 cups) baby English
    spinach leaves
50 g (1⅔ cups) watercress sprigs
200 g (7 oz) semi-dried (sun-blushed)
    tomatoes

Put all the ingredients for the horseradish cream in a small bowl. Add some freshly ground black pepper and whisk together well. Cover and refrigerate for 15 minutes, or until required.

Meanwhile, bring a saucepan of salted water to the boil. Add the beans and cook for 4 minutes, or until tender. Drain, refresh under cold water, then drain well and set aside.

Heat a chargrill pan or barbecue hotplate to high. Brush the steak slices and onion halves with the olive oil. Grill the steak slices for 2 minutes on each side, or until rare. Remove to a plate, cover loosely with foil and set aside to rest for 5 minutes.

Meanwhile, grill the onion for 2–3 minutes on each side, or until lightly charred and tender.

Toss the spinach, watercress, tomatoes and beans together in a large salad bowl.

Slice the beef thinly across the grain, then arrange over the salad. Slice the grilled onion, add to the salad and drizzle with the horseradish cream. Season well with sea salt and freshly ground black pepper and serve.

NOTE: This cooking time will result in rare beef. Cook it for a little longer if you prefer your beef medium or well done.

Pizza spinaci

MAKES TWO 30 CM (12 INCH) PIZZAS

1 tablespoon caster (superfine) sugar

2 teaspoons active dried yeast, or
    15 g ($^1/_2$ oz) fresh yeast

210 ml (7$^1/_2$ fl oz) lukewarm water

450 g (1 lb/3$^2/_3$ cups) plain (all-purpose)
    flour

$^1/_4$ teaspoon sea salt

3 tablespoons olive oil

cornflour (cornstarch), for dusting

TOPPING

4 tablespoons olive oil, plus extra,
    for brushing

4 garlic cloves, crushed

4 tablespoons pine nuts

1 kg (2 lb 4 oz) baby English spinach
    leaves

400 ml (14 fl oz) ready-made tomato
    pasta sauce

440 g (15$^1/_2$ oz/3 cups) grated mozzarella
    cheese

30 very small black olives

50 g (1$^3/_4$ oz/$^1/_2$ cup) grated parmesan
    cheese

Put the sugar and yeast in a bowl and stir in 90 ml (3 fl oz) of the lukewarm water. Leave to stand in a draught-free place for 10 minutes, or until foamy.

Mix the flour and sea salt in a large bowl and make a well in the centre. Add the olive oil, remaining lukewarm water and the yeast mixture to the well and mix using a wooden spoon until a rough dough forms. Transfer to a lightly floured surface and knead for 8 minutes, adding a little flour or extra warm water as necessary, until the dough is soft, smooth and elastic.

Place the dough in a large oiled bowl, turning to coat in the oil. Cover with plastic wrap and leave in a draught-free place for 1–1$^1/_2$ hours, or until doubled in size.

Preheat the oven to 240°C (475°F/Gas 8) and lightly oil two 30 cm (12 inch) rectangular or round baking trays.

To make the topping, heat the olive oil in a large saucepan, then add the garlic and pine nuts and fry over low heat, stirring often, for 5–6 minutes, or until golden. Add the spinach (in batches if necessary), increase the heat and stir until wilted. Season with sea salt and freshly ground black pepper and set aside.

Gently deflate the dough using a lightly floured fist, then divide in half. Roll out each portion to fit the baking trays.

Dust each pizza base with cornflour and spoon half the tomato sauce onto each base, spreading nearly to the edges. Sprinkle with half the mozzarella. Spread the spinach mixture and olives over the top, then sprinkle with the remaining mozzarella and parmesan.

Bake for 12–15 minutes, or until the crust is golden and puffed. Brush the rim with a little extra olive oil before serving.

NOTE: This recipe will serve 12 people, but can easily be halved to serve six.

SPINACH, RICE AND BACON PIE

SERVES 4–6

750 ml (26 fl oz/3 cups) beef stock
45 g (1½ oz) unsalted butter
2 tablespoons olive oil
1 large onion, finely chopped
2 garlic cloves, finely chopped
175 g (6 oz) bacon slices, trimmed and
  chopped
220 g (7¾ oz/1 cup) risotto rice
800 g (1 lb 10 oz) English spinach,
  trimmed and coarsely chopped
4 eggs, lightly beaten
50 g (1¾ oz/½ cup) grated parmesan
  cheese
1 teaspoon coarsely cracked black
  pepper
4 tablespoons dry breadcrumbs

Pour the stock into a saucepan and bring to the boil.
Reduce the heat, then cover and keep at simmering point.

Heat the butter and half the olive oil in a large
heavy-based frying pan. Add the onion and cook over
medium heat for 3–4 minutes, then add the garlic and
bacon and cook for 1 minute.

Add the rice and stir to coat, then pour in 125 ml
(4 fl oz/½ cup) of the simmering stock and cook over
low heat, stirring constantly, until all the stock has been
absorbed. Continue adding the stock, 125 ml (4 fl oz/½ cup)
at a time, stirring constantly and making sure the stock has
been absorbed before adding more. Cook until all the stock
is absorbed.

Meanwhile, preheat the oven to 180°C (350°F/Gas 4).

Stir the spinach into the rice, then cover and simmer
for 2 minutes, or until just wilted. Transfer to a bowl and leave
to cool a little. Stir in the eggs, parmesan and pepper.

Grease a 23 cm (9 inch) spring-form tin and sprinkle
with 3 tablespoons of the breadcrumbs. Spoon the filling in,
drizzle with the remaining olive oil and sprinkle the remaining
breadcrumbs over the top.

Bake for 40–45 minutes, then remove from the oven
and allow to cool in the tin. Cut into wedges and serve at
room temperature.

SPANAKORIZO
SERVES 6 AS A SIDE DISH OR LIGHT MEAL

400 g (14 oz) English spinach (see Note)
6 spring onions (scallions)
2 tablespoons olive oil
1 large onion, chopped
2 garlic cloves, crushed
330 g (11 1/2 oz/1 1/2 cups) white rice
  (use a short or medium-grain)
1 tablespoon chopped dill
1 tablespoon chopped flat-leaf (Italian)
  parsley
2 tablespoons lemon juice
375 ml (13 fl oz/1 1/2 cups) vegetable
  stock

Wash the spinach, drain well, then tear the leaves and chop the stalks. Finely chop the spring onions, including the green tops.

Heat the olive oil in a large heavy-based saucepan with a tight-fitting lid. Add the onion and sauté over medium heat for 5 minutes, or until softened. Stir in the garlic and cook for a further minute.

Add the spring onion and rice and cook for 2 minutes, stirring constantly to coat the rice. Add the spinach, herbs and half the lemon juice. Season well with sea salt and freshly ground black pepper.

Stir in the stock and 375 ml (13 fl oz/1 1/2 cups) water. Cover, bring to the boil, then reduce the heat and simmer for 15 minutes.

Remove from the heat and set aside, covered, for 5 minutes. Stir in the remaining lemon juice, then adjust the seasoning if needed and serve.

NOTE: Silverbeet (Swiss chard) can be used instead of English spinach. Rinse the silverbeet, then cut off the thick stems and roughly chop the leaves. Blanch the leaves in a large saucepan of boiling salted water. Rinse under cold water and add to the rice with the herbs.

Warm silverbeet and chickpea salad with sumac

Serves 4 as a side dish

220 g (7 ¾ oz/1 cup) dried chickpeas
125 ml (4 fl oz/½ cup) olive oil
1 onion, cut into thin wedges
2 tomatoes
1 teaspoon sugar
¼ teaspoon ground cinnamon
2 garlic cloves, chopped
1.5 kg (3 lb 5 oz) silverbeet (Swiss chard)
3 tablespoons chopped mint
2–3 tablespoons lemon juice
1½ tablespoons ground sumac (see Note)

Put the chickpeas in a large bowl, cover with plenty of cold water and leave to soak overnight.

Drain the chickpeas and place in a large saucepan. Cover with plenty of cold fresh water and bring to the boil, then reduce the heat and simmer for 1¾ hours, or until tender. Drain well and set aside.

Heat the olive oil in a saucepan. Add the onion and sauté over low heat for 5 minutes, or until softened and just starting to brown.

Meanwhile, cut the tomatoes in half, scrape out the seeds with a teaspoon and chop the flesh. Add the tomato to the pan with the sugar, cinnamon and garlic and cook for 2–3 minutes, or until softened.

Remove the stems and tough ribs from the silverbeet, then finely slice the leaves and add to the pan with the chickpeas. Cook for 3–4 minutes, or until the silverbeet wilts. Stir in the mint, lemon juice and sumac and season to taste with sea salt and freshly ground black pepper. Cook for a final minute and serve.

NOTE: Sumac is a peppery spice widely used in Middle Eastern cookery. It is available from most gourmet food stores.

GREEN BEANS WITH GARLIC BREADCRUMBS
SERVES 4 AS A SIDE DISH

600 g (1 lb 5 oz) baby green beans,
   trimmed
3 tablespoons olive oil
4 garlic cloves
40 g (1½ oz/½ cup) fresh breadcrumbs
2 tablespoons chopped flat-leaf (Italian)
   parsley

Bring a large saucepan of salted water to the boil. Add the beans and cook for 3 minutes, or until tender but still firm. Drain, refresh under cold water, then drain well and pat dry with paper towels.

Heat the olive oil in a large frying pan. Add the garlic cloves and cook over medium heat for 5–6 minutes, or until golden brown. Remove the garlic using a slotted spoon and discard.

Add the breadcrumbs to the pan and cook over low heat, stirring constantly, for 3–4 minutes, or until brown and crunchy. Add the beans and parsley, mix together and briefly cook to warm the beans. Season with sea salt and freshly ground black pepper and serve warm or at room temperature.

GREEN BEANS WITH TOMATO AND OLIVE OIL
SERVES 4 AS A SIDE DISH

4 tablespoons olive oil
1 large onion, chopped
3 garlic cloves, finely chopped
400 g (14 oz) tin chopped tomatoes
½ teaspoon sugar
750 g (1 lb 10 oz) green beans, trimmed
3 tablespoons chopped flat-leaf (Italian)
   parsley

Heat the olive oil in a large frying pan. Add the onion and sauté over medium heat for 5 minutes, or until softened. Add the garlic and cook for a further 30 seconds.

Add the tomatoes, sugar and 125 ml (4 fl oz/½ cup) water, then season with sea salt and freshly ground black pepper. Bring to the boil, reduce the heat and simmer for 10 minutes, or until the liquid has reduced slightly.

Add the beans, then partially cover and simmer for a further 10 minutes, or until the beans are tender and the tomato mixture is pulpy.

Stir in the parsley, check the seasoning and serve.

CHARGRILLED VEGETABLE TERRINE

SERVES 8 AS A STARTER

8 large slices of chargrilled eggplant
(aubergine), drained
10 slices of chargrilled red capsicum
(pepper), drained
8 slices of chargrilled zucchini
(courgette), drained
350 g (12 oz) ricotta cheese
2 garlic cloves, crushed
45 g (1½ oz) rocket (arugula) leaves
3 marinated artichokes, drained and
sliced
85 g (3 oz/heaped ½ cup) semi-dried
(sun-blushed) tomatoes, drained
and chopped
100 g (3½ oz) marinated mushrooms,
drained and halved

Line a 23 x 13 x 6 cm (9 x 5 x 2½ inch) loaf (bar) tin with plastic wrap, leaving a generous overhang on each side.

Line the base of the tin with half the eggplant, cutting it and patching it to fit evenly. Lay half the capsicum evenly over the eggplant, then layer all the zucchini evenly over the top.

Using a wooden spoon, beat the ricotta and garlic in a bowl until smooth. Season to taste with sea salt and freshly ground black pepper, then spread the mixture evenly over the zucchini, pressing down firmly. Top with the rocket leaves. Arrange the artichoke, tomato and mushroom over the rocket.

Top with another layer of capsicum, then finish with the remaining eggplant.

Cover the terrine tightly, using the overhanging plastic wrap. Place a piece of cardboard on top of the terrine, then use tins of food to press it down. Refrigerate overnight.

To serve, carefully peel back the plastic wrap and turn the terrine out onto a platter. Remove the plastic wrap, cut into thick slices and serve.

NICOISE SALAD WITH GREEN BEANS AND SEARED TUNA
SERVES 4

DRESSING
4 tablespoons olive oil
2 garlic cloves, crushed
1 teaspoon dijon mustard
1 tablespoon Champagne vinegar

2 x 150 g (5½ oz) tuna fillets
olive oil, for brushing
225 g (8 oz) green beans, trimmed
450 g (1 lb) new potatoes, scrubbed
    and halved
2 large handfuls of torn butter lettuce
    leaves
8 cherry tomatoes, halved
12 kalamata olives
2 eggs, hard-boiled
4 anchovies, halved
1 tablespoon salted capers, rinsed
    and drained

In a small bowl, whisk together the dressing ingredients. Season well with sea salt and freshly ground black pepper and set aside.

Preheat a barbecue grill plate or a large heavy-based frying pan to medium–high. Brush the tuna fillets with olive oil and cook for 1½–2 minutes on each side, or until well browned but still a little rare in the middle. Transfer to a plate and set aside to cool.

Bring a saucepan of salted water to the boil. Add the beans and cook for 3 minutes, or until just tender. Remove with tongs and drain. Add the potatoes to the saucepan and boil for 10 minutes, or until tender. Drain, allow to cool slightly, then place in a shallow salad bowl with the lettuce leaves. Add 2 tablespoons of the dressing and toss gently to coat. Scatter the tomatoes over the top.

Slice the tuna on the diagonal into 5 mm (¼ inch) slices and arrange over the tomatoes. Add the beans and olives. Shell the hard-boiled eggs, cut into wedges and add to the salad. Scatter the anchovies and capers over the top, drizzle with the remaining dressing and serve.

BRAISED CELERY
SERVES 4 AS A SIDE DISH

30 g (1 oz) unsalted butter
1 bunch of celery, trimmed and cut into
　5 cm (2 inch) lengths
500 ml (17 fl oz/2 cups) chicken or
　vegetable stock
2 teaspoons finely grated lemon zest
3 tablespoons lemon juice
3 tablespoons pouring (whipping) cream
2 egg yolks
1 tablespoon cornflour (cornstarch)
a pinch of ground mace or nutmeg
1–2 tablespoons chopped parsley

Preheat the oven to 180°C (350°F/Gas 4). Grease a large shallow baking dish.

Melt the butter in a large frying pan. Add the celery, toss to coat evenly in the butter, then cover and cook over medium heat for 2 minutes.

Pour in the stock. Add the lemon zest and lemon juice, then cover and simmer for 10 minutes, or until the celery is tender, but still holds its shape. Remove the celery using a slotted spoon and place in the baking dish. Reserve 3 tablespoons of the cooking liquid.

In a bowl, mix together the cream, egg yolks and cornflour. Whisk in the reserved cooking liquid. Pour the mixture back into the frying pan and cook, stirring constantly, until the mixture boils and thickens. Add the mace and season to taste with sea salt and freshly ground black pepper.

Pour the sauce over the celery and bake for 15 minutes, or until the celery is very soft and the sauce is bubbling. Scatter the parsley over the top.

Serve warm with poached chicken breast, chargrilled lamb or slices of corned beef.

CUCUMBER, FETA, MINT AND DILL SALAD
SERVES 4

125 g (4½ oz) feta cheese
4 Lebanese (short) cucumbers
1 small red onion, thinly sliced
1½ tablespoons finely chopped dill
1 tablespoon dried mint
3 tablespoons olive oil
1½ tablespoons lemon juice
crusty bread, to serve

Crumble the feta into 1 cm (½ inch) chunks and place in a large bowl. Cut the cucumbers into 1 cm (½ inch) lengths, then add to the bowl along with the onion and dill.

Grind the mint in a mortar and pestle, or force it through a sieve until powdered. Tip into a small bowl, add the olive oil and lemon juice and whisk until combined. Season with a little sea salt and freshly ground black pepper, pour over the salad and toss well. Serve with crusty bread.

WALDORF SALAD
SERVES 4

butter lettuce leaves, to serve
2 red apples, quartered and cored
1 large green apple, quartered and cored
1½ celery stalks, sliced
3 tablespoons walnut halves
2 tablespoons ready-made egg
    mayonnaise
1 tablespoon sour cream

Line a serving bowl with lettuce leaves. Cut all the apples into 2 cm (¾ inch) chunks and place in a large mixing bowl with the celery and walnuts.

In a small bowl, mix together the mayonnaise and sour cream. Stir the dressing through the salad mixture, then transfer to the lettuce-lined serving bowl and serve.

Cucumber, feta, mint and dill salad

BARBECUED SALMON CUTLETS WITH SWEET CUCUMBER DRESSING
SERVES 4

SWEET CUCUMBER DRESSING
2 small Lebanese (short) cucumbers,
   peeled, seeded and finely diced
1 red onion, finely chopped
1 red chilli, finely chopped
2 tablespoons pickled ginger, shredded
   (see Note)
2 tablespoons rice vinegar
1/2 teaspoon sesame oil

oil, for brushing
4 salmon (or ocean trout) cutlets
1 sheet of toasted nori (dried seaweed),
   cut into thin strips (see Note)
steamed rice, to serve

In a bowl, mix together all the ingredients for the sweet cucumber dressing. Cover with plastic wrap and leave to stand at room temperature while cooking the salmon.

Heat a barbecue hotplate or large heavy-based frying pan to medium–high, then lightly brush with oil. Cook the salmon cutlets for 2 minutes on each side, or until cooked but still a little pink in the middle — take care not to overcook the fish or it will be dry.

Transfer the salmon to warmed serving plates. Spoon the sweet cucumber dressing over and sprinkle with toasted nori strips. Serve with steamed rice.

NOTE: Pickled ginger and nori are Japanese ingredients available from Asian food stores and larger supermarkets.

LOBSTER SOUP WITH ZUCCHINI AND AVOCADO

SERVES 4

50 g (1³/₄ oz) unsalted butter
2 shallots, finely chopped
1 onion, finely chopped
1 zucchini (courgette), trimmed and cut
   into 5 mm (¹/₄ inch) pieces
1 garlic clove, crushed
2¹/₂ tablespoons dry white wine
400 ml (14 fl oz) fish stock
250 g (9 oz) raw lobster meat, chopped
250 ml (9 fl oz/1 cup) thick (double/
   heavy) cream
1 avocado, diced
1 tablespoon chopped coriander
   (cilantro) leaves
1 tablespoon chopped parsley
lemon juice, to serve

Melt the butter in a large saucepan. Add the shallot, onion, zucchini and garlic. Sauté over medium heat for 5–6 minutes, or until the vegetables are just soft.

Add the wine and bring to the boil, then cook for 3 minutes. Pour in the stock and bring back to the boil. Reduce the heat to low, add the lobster and simmer for 3–4 minutes, or until the lobster meat is opaque. Stir in the cream and season to taste with sea salt and freshly ground black pepper.

Divide the soup among four warmed bowls, then stir some of the avocado, coriander and parsley into each. Drizzle with a little lemon juice and serve.

VARIATION: Crayfish or prawns (shrimp) can be substituted for the lobster.

ZUCCHINI PATTIES
SERVES 4 AS A STARTER OR SIDE DISH

CUCUMBER AND YOGHURT SALAD
1 Lebanese (short) cucumber
sea salt, for sprinkling
250 g (9 oz/1 cup) Greek-style yoghurt
1 small garlic clove, crushed
1 tablespoon chopped dill
2 teaspoons white wine vinegar
ground white pepper, to taste

300 g (10 oz) zucchini (courgettes),
    grated
1 small onion, finely chopped
3 tablespoons self-raising flour
4 tablespoons grated kefalotyri or
    parmesan cheese
1 tablespoon chopped mint
2 teaspoons chopped flat-leaf (Italian)
    parsley
a pinch of ground nutmeg
3 tablespoons dry breadcrumbs
1 egg, lightly beaten
olive oil, for pan-frying
rocket (arugula) leaves, to serve
lemon wedges, to serve (optional)

To make the cucumber and yoghurt salad, chop the cucumber into small pieces, place in a colander, sprinkle with sea salt and set aside in the sink or on a plate to drain for 15–20 minutes.

In a bowl, mix together the yoghurt, garlic, dill and vinegar. Add the cucumber and season to taste with sea salt and ground white pepper. Cover and refrigerate until required.

Meanwhile, preheat the oven to 120°C (235°F/Gas ½).

Put the zucchini and onion in a clean tea towel (dish towel), gather the corners together and twist as tightly as possible to remove all the juices. Tip the zucchini and onion into a large bowl, then add the flour, cheese, mint, parsley, nutmeg, breadcrumbs and egg. Season well with sea salt and freshly cracked black pepper, then mix with your hands to a stiff batter.

Heat 1 cm (½ inch) olive oil in a large heavy-based frying pan over medium heat. When the oil is hot, drop 2 tablespoons of the batter into the pan and press flat to make a thick patty. Fry several at a time for 2–3 minutes, or until well browned all over. Drain well on paper towels and place in the oven to keep warm while cooking the remaining patties.

Serve hot with rocket leaves and the cucumber and yoghurt salad, and perhaps some lemon wedges.

BRUSSELS SPROUTS WITH PANCETTA

SERVES 4 AS A SIDE DISH

100 g (3½ oz) pancetta, thinly sliced
4 shallots
20 g (¾ oz) butter
1 tablespoon olive oil
1 garlic clove, crushed
500 g (1 lb 2 oz) brussels sprouts,
   trimmed and thickly sliced

Preheat the oven grill (broiler) to high. Spread the pancetta on a baking tray lined with foil and place 8–10 cm (3¼–4 inches) under the heat. Grill for 1 minute, or until crisp, then set aside to cool.

Put the shallots in a saucepan of boiling water for 5 minutes to make them easier to peel. Remove the shallots using a slotted spoon, allow to cool slightly, then peel and cut into thick rings.

Heat the butter and olive oil in a large frying pan. Add the shallot and garlic and sauté over medium heat for 3–4 minutes, or until just starting to brown. Add the brussels sprouts and season with freshly ground black pepper. Sauté for 4–5 minutes, or until the brussels sprouts are light golden and crisp. Turn off the heat, cover and set aside for 5 minutes.

Break the pancetta into large pieces, gently toss through the vegetables and serve.

PRAWN AND FENNEL SALAD
SERVES 4

1.25 kg (2 lb 12 oz) raw large prawns
  (shrimp)
300 g (10½ oz) watercress
1 large fennel bulb, thinly sliced
2 tablespoons finely snipped chives
125 ml (4 fl oz/½ cup) extra virgin
  olive oil
3 tablespoons lemon juice
1 tablespoon dijon mustard
1 large garlic clove, finely chopped

Peel the prawns and remove the veins. Meanwhile, bring a saucepan of water to the boil. Add the prawns to the saucepan, return to the boil and simmer for 2 minutes, or until the prawns turn opaque and are just cooked through. Drain well and leave to cool, then slice in half lengthways. Place in a large serving bowl.

Pick the sprigs from the watercress, discarding the stalks. Wash the sprigs and dry well, then add to the prawns with the fennel and chives. Mix well.

In a bowl, whisk together the olive oil, lemon juice, mustard and garlic to make a dressing. Pour the dressing over the salad, season with sea salt and freshly cracked black pepper and toss gently. Arrange on serving plates and serve.

FLORENTINE ROAST PORK

SERVES 6

3 large fennel bulbs, with fronds
1/2 tablespoon finely chopped rosemary
4 garlic cloves, crushed
1.5 kg (3 lb 5 oz) pork loin, chined and
  skinned (see Note)
3 white onions
90 ml (3 fl oz) olive oil
185 ml (6 fl oz/3/4 cup) dry white wine
4 tablespoons extra virgin olive oil
250 ml (9 fl oz/1 cup) chicken stock
3–4 tablespoons thick (double/heavy)
  cream

Preheat the oven to 200°C (400°F/Gas 6).

Cut the green fronds from the very tops of the fennel and chop to give 2 tablespoons fronds. Place in a small bowl with the rosemary, garlic and some sea salt and freshly ground black pepper. Using a small sharp knife, make deep incisions all over the pork and rub the fennel mixture into the incisions.

Cut two of the onions in half and place in a flameproof roasting tin. Sit the pork on top of the onion and drizzle the olive oil over the top.

Roast the pork for 30 minutes, then baste with the pan juices and reduce the heat to 180°C (350°F/Gas 4). Roast for a further 30 minutes, then baste and lightly salt the surface of the pork. Pour half the wine into the pan and roast for a further 30–45 minutes, or until the pork is cooked, basting once or twice. To test if the meat is cooked, insert a skewer in the thickest part — the juices should run clear.

Meanwhile, remove the tough outer layers from the fennel and discard. Slice the bulbs lengthways into slices 1 cm (1/2 inch) thick and place in a large saucepan. Thinly slice the remaining onion and add to the saucepan with the extra virgin olive oil and a little sea salt. Add enough water to cover the vegetables, then cover the pan and bring to the boil. Reduce the heat and simmer for 45 minutes, or until the fennel is soft. Drain the pan, if necessary, and keep warm.

Remove the pork from the roasting tin, cover loosely with foil and leave to rest in a warm place for 15–20 minutes. Skim the excess oil from the pan juices in the roasting tin, discarding the onion.

Place the roasting tin on the stovetop over high heat. Add the remaining wine, stirring to loosen any bits stuck to the bottom of the tin. Add the stock and boil until the sauce has slightly thickened. Remove from the heat, season with sea salt and freshly ground black pepper and stir in the cream.

Divide the fennel among warmed plates. Slice the pork and serve over the fennel, with the sauce passed separately.

NOTE: Chining involves cutting through the tip of the vertebrae within the joint to make carving easier. Ask your butcher to do this for you.

FENNEL CRUMBLE

SERVES 6 AS A SIDE DISH

100 ml (3¹/₂ fl oz) lemon juice
2 fennel bulbs, or 5 baby fennel bulbs
1 tablespoon honey
1 tablespoon plain (all-purpose) flour
310 ml (10³/₄ fl oz/1¹/₄ cups) pouring
  (whipping) cream

CRUMBLE TOPPING
75 g (2¹/₂ oz/³/₄ cup) rolled (porridge)
  oats
60 g (2¹/₄ oz/¹/₂ cup) plain (all-purpose)
  flour
110 g (3³/₄ oz/1 cup) fresh black rye
  breadcrumbs (see Note)
60 g (2¹/₄ oz) unsalted butter
1 garlic clove, crushed

Preheat the oven to 180°C (350°F/Gas 4). Grease a large heatproof serving dish.

Bring a large saucepan of water to the boil and add 3 tablespoons of the lemon juice. Trim the fennel and cut into thin slices. Wash and drain well, then add to the boiling water and cook over medium heat for 3 minutes. Drain well and allow to cool slightly.

Put the fennel in a large bowl. Add the honey and remaining lemon juice and season with freshly ground black pepper. Sprinkle with the flour and toss to combine. Spoon into the prepared dish and pour the cream over the top.

To make the crumble topping, put the oats, flour and breadcrumbs in a bowl. Melt the butter in a small saucepan, add the garlic and cook for 30 seconds. Pour over the dry ingredients and mix well.

Sprinkle the crumble topping over the fennel. Bake for 20–30 minutes, or until the fennel is tender and the topping is nicely browned. Serve hot.

NOTE: White or wholemeal (whole-wheat) breadcrumbs can be used in place of rye bread, if preferred.

SPAGHETTI WITH SARDINES, FENNEL AND TOMATO
SERVES 4–6

3 roma (plum) tomatoes
4 tablespoons olive oil
80 g (2¾ oz/1 cup) fresh white
  breadcrumbs
3 garlic cloves, crushed
1 red onion, thinly sliced
1 fennel bulb, quartered and thinly sliced
3 tablespoons raisins
3 tablespoons pine nuts, toasted
4 anchovy fillets, chopped
125 ml (4 fl oz/½ cup) white wine
1 tablespoon tomato paste (concentrated
  purée)
4 tablespoons finely chopped flat-leaf
  (Italian) parsley
350 g (12 oz) butterflied sardines (ask
  your fishmonger to do this)
500 g (1 lb 2 oz) spaghetti

Bring a saucepan of water to the boil. Using a small sharp knife, score a small cross in the base of each tomato. Place the tomatoes in the boiling water for about 20 seconds, remove using a slotted spoon, then plunge into a bowl of iced water. Drain the tomatoes and peel the skins away from the cross. Cut the tomatoes in half and scoop out the seeds with a teaspoon, then roughly chop the flesh.

Heat 1 tablespoon of the olive oil in a large frying pan over medium heat. Add the breadcrumbs and one-third of the garlic and cook, stirring often, for 5 minutes, or until the breadcrumbs are golden and crisp. Remove from the pan using a slotted spoon.

Heat the remaining oil in the same pan. Add the onion, fennel and remaining garlic and sauté for 8 minutes, or until the vegetables are soft. Add the chopped tomato, raisins, pine nuts and anchovies and cook for a further 3 minutes, then stir in the wine, tomato paste and 125 ml (4 fl oz/½ cup) water. Simmer for 10 minutes, or until the mixture thickens slightly. Stir in the parsley, then set aside and keep warm.

Pat the sardines dry with paper towels. Cook the sardines in batches in a lightly greased frying pan over medium heat for 1–2 minutes, or until just cooked through. Take care not to overcook or they will break up. Set aside.

Meanwhile, cook the spaghetti in a large saucepan of rapidly boiling salted water until *al dente*. Drain and return to the pan.

Add the sauce to the spaghetti and stir to coat well, then add the sardines and half the fried breadcrumbs and toss gently to combine. Sprinkle the remaining breadcrumbs over the top and serve immediately.

BEETROOT HUMMUS
SERVES 8

500 g (1 lb 2 oz) beetroot (beets)
4 tablespoons olive oil
1 large onion, chopped
1 tablespoon ground cumin
400 g (14 oz) tin chickpeas, drained
1 tablespoon tahini
4 tablespoons plain yoghurt
3 garlic cloves, crushed
3 tablespoons lemon juice
125 ml (4 fl oz/½ cup) vegetable stock
pitta or Turkish bread, to serve

Scrub the beetroot well. Bring a large saucepan of salted water to the boil, add the beetroot and simmer for 30–45 minutes, or until tender when pierced with a skewer. Drain well and leave until cool enough to handle. Peel away the skin, then roughly chop the flesh and set aside.

Meanwhile, heat 1 tablespoon of the olive oil in a frying pan. Add the onion and sauté over medium heat for 5 minutes, or until soft. Add the cumin and cook for 1 minute more, or until fragrant.

Chop the beetroot and place in a food processor with the onion mixture, chickpeas, tahini, yoghurt, garlic, lemon juice and stock. Process until smooth. With the motor running, add the remaining olive oil in a thin, steady stream until thoroughly combined. Serve with pitta or Turkish bread.

NOTE: You can use 500 g (1 lb 2 oz) of any cooked vegetable to make this hummus. Try carrot or pumpkin (winter squash).

ROASTED BEETROOT WITH HORSERADISH CREAM
SERVES 4–6 AS A SIDE DISH

8 beetroot (beets), scrubbed
2 tablespoons olive oil
2 teaspoons honey
75 ml (2¼ fl oz) thick (double/heavy) cream
1 tablespoon grated fresh horseradish
1 teaspoon lemon juice
a small pinch of sugar
chopped parsley or dill, to garnish

Preheat the oven to 200°C (400°F/Gas 6). Peel the beetroot, wearing gloves to stop your hands staining. Trim the ends, cut into quarters and divide among four large squares of foil.

Put the olive oil and honey in a small bowl, season with sea salt and freshly ground black pepper and mix well. Drizzle over the beetroot, turning to coat well. Wrap the foil loosely around the beetroot and bake for 1–1½ hours, or until tender when pierced with a skewer. Remove from the oven and leave the beetroot in the foil for 5 minutes to cool slightly.

Whip the cream until starting to thicken, then fold in the horseradish, lemon juice, sugar and a pinch of sea salt.

Serve the warm beetroot with a generous dollop of the horseradish cream, sprinkled with chopped parsley or dill.

BEETROOT WITH SKORDALIA
SERVES 6 AS A STARTER OR SIDE DISH

1 kg (2 lb 4 oz) beetroot (beets), with
  leaves attached
3 tablespoons extra virgin olive oil
1 tablespoon red wine vinegar

SKORDALIA
250 g (9 oz) roasting potatoes, such as
  russet (idaho) or king edward, peeled
  and cut into 2 cm (^{3}/$_4$ inch) cubes
2–3 garlic cloves, crushed
^{1}/$_2$ teaspoon sea salt
ground white pepper, to taste
90 ml (3 fl oz) olive oil
1 tablespoon white vinegar

Cut the stems from the beetroot, leaving 2–3 cm (^{3}/$_4$–1^1/$_4$ inches)
attached. Wash the leaves, discarding any tough outer ones.
Cut the stems and leaves into 7 cm (2^3/$_4$ inch) lengths and
wash well. Scrub the beetroot bulbs clean.

Bring a large saucepan of salted water to the boil. Add
the beetroot bulbs and gently boil for 30–45 minutes, or until
tender when pierced with a skewer. Remove with a slotted
spoon and cool slightly.

Meanwhile, make the skordalia. Bring another large
saucepan of water to the boil, add the potato and cook for
10 minutes, or until very soft. Drain thoroughly, then mash
using a masher or potato ricer until quite smooth. Stir in the
garlic, sea salt and a pinch of white pepper, then gradually add
the olive oil, stirring well with a wooden spoon. Stir in the
vinegar and season to taste.

Bring the beetroot water back to the boil. Add the leaves
(and a little more water if necessary) and boil for 8 minutes, or
until tender. Drain well, allow to cool slightly, then squeeze
out any excess water from the leaves.

Peel the beetroot, then cut into quarters or thick wedges.
Arrange on a serving plate with the leaves. Mix together the
extra virgin olive oil and vinegar, season to taste and drizzle
over the leaves and bulbs. Serve warm or at room temperature,
with the skordalia.

LEEKS À LA GRECQUE
SERVES 4 AS A SIDE DISH OR STARTER

3 tablespoons extra virgin olive oil
1½ tablespoons white wine
1 tablespoon tomato paste
   (concentrated purée)
¼ teaspoon sugar
1 bay leaf
1 thyme sprig
1 garlic clove, crushed
4 coriander seeds, crushed
4 peppercorns
8 small leeks, white part only, rinsed well
1 teaspoon lemon juice
1 tablespoon chopped parsley
lemon halves or wedges, to serve

Put the olive oil, wine, tomato paste, sugar, bay leaf, thyme, garlic, coriander seeds, peppercorns and 250 ml (9 fl oz/1 cup) water in a large heavy-based frying pan with a lid. Bring to the boil, cover and simmer for 5 minutes.

Add the leeks in a single layer and bring to simmering point. Reduce the heat, then cover and simmer gently for 20–30 minutes, or until the leeks are tender when pierced with a skewer. Drain the leeks well, reserving the liquid, then transfer to a serving dish.

Add the lemon juice to the reserved cooking liquid and boil rapidly for 1 minute, or until the liquid has reduced and is slightly syrupy. Season to taste with sea salt, then strain the sauce over the leeks.

Allow to cool, then serve the leeks at room temperature, sprinkled with chopped parsley and with some lemon for squeezing over.

BRAISED LEEK WITH PINE NUTS
SERVES 4 AS A SIDE DISH

20 g (¾ oz) unsalted butter
2 teaspoons olive oil
2 leeks, white part only, thinly sliced
4 tablespoons vegetable stock
4 tablespoons dry white wine
2 tablespoons finely chopped mixed
   herbs, such as flat-leaf (Italian) parsley
   and oregano
2½ tablespoons pine nuts, lightly toasted
4 tablespoons grated parmesan cheese

Heat the butter and olive oil in a large frying pan. Add the leek and sauté for 5 minutes, or until golden brown.

Add the stock and wine and cook for a further 10 minutes, or until the leek is tender.

Stir in the herbs, sprinkle with the pine nuts and parmesan and serve.

FARMHOUSE RHUBARB PIE
SERVES 6

185 g (6½ oz/1½ cups) plain
   (all-purpose) flour
2 tablespoons icing (confectioners')
   sugar
125 g (4½ oz) cold unsalted butter,
   chopped
1 egg yolk, mixed with 1 tablespoon
   iced water

FILLING
220 g (7¾ oz/1 cup) sugar, plus extra,
   for sprinkling
750 g (1 lb 10 oz/6 cups) chopped
   rhubarb
2 large apples, peeled, cored and
   chopped
2 teaspoons grated lemon zest
3 pieces of preserved ginger, sliced
ground cinnamon, for sprinkling
icing (confectioners') sugar, for dusting
   (optional)

Sift the flour, icing sugar and a pinch of sea salt into a large
bowl. Using your fingertips, lightly rub the butter into the
flour until the mixture resembles coarse breadcrumbs. Make
a well in the centre. Add the egg yolk mixture to the well
and mix using a flat-bladed knife until a rough dough forms.
Gently gather the dough together, transfer to a lightly floured
surface, then press into a round disc. Cover with plastic wrap
and refrigerate for 30 minutes, or until firm.

Meanwhile, preheat the oven to 190°C (375°F/Gas 5).
Grease a 20 cm (8 inch) pie plate.

Roll the pastry out to a 35 cm (14 inch) circle and ease it
into the pie plate, allowing the excess to hang over the edge.
Refrigerate the pastry-lined dish while preparing the filling.

In a saucepan, heat the sugar and 125 ml (4 fl oz/½ cup)
water for 4–5 minutes, or until syrupy. Add the rhubarb, apple,
lemon zest and ginger. Cover and gently simmer for 5 minutes,
or until the rhubarb is cooked but still holds its shape.

Drain off the liquid and allow the rhubarb to cool. Spoon
into the pastry shell and sprinkle with the cinnamon and a
little extra sugar. Fold the overhanging pastry over the filling
and bake for 40 minutes, or until golden. Dust with icing sugar
before serving, if desired.

RHUBARB YOGHURT CAKE
SERVES 8

310 g (11 oz/2½ cups) self-raising flour
150 g (5½ oz/1¼ cups) finely sliced
   rhubarb
230 g (8 oz/1 cup) caster (superfine) sugar
1 teaspoon natural vanilla extract
2 eggs, lightly beaten
125 g (4 oz/½ cup) plain yoghurt,
   plus extra, to serve
1 tablespoon rosewater
125 g (4½ oz) unsalted butter, melted

Preheat the oven to 180°C (350°F/Gas 4). Lightly grease a 23 cm
(9 inch) round cake tin and line the base with baking paper.

Sift the flour into a bowl, then stir in the rhubarb and sugar.
Add the remaining ingredients, stirring until just combined.

Spoon the batter into the prepared cake tin and bake
for 1 hour, or until a cake tester inserted into the centre of
the cake comes out clean. Remove from the oven and leave to
stand in the tin for 15 minutes, before turning out onto a wire
rack to cool completely. Serve with yoghurt, if desired.

Farmhouse rhubarb pie

fruit bowl

Nearly every home has a fruit bowl, that colourful repository of apples, oranges, pears and tropical fruits. The fruit bowl is also where avocados and tomatoes need to ripen, and where capsicums (peppers) and eggplants (aubergine) will keep beautifully for a day or two before cooking. Whatever the season, always try to have yours full of something!

APPLES

Apples belong to the rose family. There are over 8000 varieties, but only relatively few are grown commercially, mainly for their uniform colour, size and shape, disease resistance and ability to travel. The peak apple season is autumn through winter, so be suspicious of apples appearing in summer. Apples do keep well, but after extended periods in cold storage suffer drastic loss in flavour and crispness, so out of season it can be hard to find apples that taste as they should. Organic apples are often your best bet.

Granny smith is a large green-skinned apple with white, tart flesh that is good for cooking and eating raw, but not so great for recipes requiring it to hold its shape, as it turns mushy. **Braeburn** is a medium-large, full-flavoured apple with red-blushed skin and juicy, intensely sweet–tart flesh best suited to eating raw. **Golden delicious** has yellowish skin and mellow, sweet, creamy flesh. Ideal for eating raw, it also holds its shape perfectly when cooked. The legendary **Cox's orange pippin** is a small round apple with greenish yellow skin blushed with russet. The flesh is creamy, crisp, juicy and firm, with full sweet–acid flavour. This apple is also suitable for cooking, but so delicious raw that it probably rarely finds its way into a pie. **Fuji** apples have pink skins with gold highlights, a light, crisp, juicy interior and sweet, refreshing flavour. They hold their shape well when cooked, but their flavour and texture are best appreciated raw. A good multi-purpose apple, **pink lady** has pretty pink skin, sweet white juicy flesh and a good sugar–acid balance. **Red delicious** has a deep red skin and slightly elongated shape. Not a good cooking apple, its crisp, white, juicy flesh is best enjoyed raw.

Look for very firm apples with their stalks attached. They should be heavy for their size, with tight, dry skin. Apples continue to ripen once picked, so the fruit bowl is a great place if yours are slightly under-ripe. Fully ripened apples will keep at cool room temperature for several days, or in a well-ventilated plastic bag in the crisper for up to 1 week.

If peeling or chopping apples ahead of time, or in a large quantity, put them in a bowl of acidulated water as you go so they don't discolour.

Apples pair well with pork, horseradish, beetroot (beet), celery, potato, sweet potato, parsnip, chicken, cheddar cheese, cream, walnuts, sage, mint, thyme, pears, rhubarb, blackberries, quince, raspberries, prunes, raisins, dates, cloves, cinnamon, nutmeg, cider, brandy, honey, brown sugar and maple syrup.

Avocado

What other fruit, instead of storing energy as sugar, stores it as nutritious, luscious, unsaturated fat? With its buttery green flesh, the avocado is doubly odd as it only ripens off the tree. If left on the tree, mature fruits will remain hard for up to 7 months without starting to spoil. High in vitamins B6, C and E, folate, potassium and fibre, the avocado is also incredibly nutritious — in its native tropics it is known as 'poor man's butter'.

The trick with an avocado is to know when it is ripe: once you cut into one, it won't ripen any further. To choose a perfectly ripe avocado, cradle it in your hands and gently apply slight pressure to the skin at the neck end — it should 'give' a little. Inspect skins closely for any signs of bruising. You can also buy your avocados hard and ripen them at room temperature, which may take 3–5 days, depending on their hardness. To speed things up, store them in a paper bag with an apple or two — the ethylene gas produced by the apples hastens the ripening process. Once ripe, refrigerate your avocados so they don't over-ripen.

The many varieties each differ a little in skin colour, texture and thickness, but all have a large, hard seed that requires removal, along with the skin. Avocado varieties are interchangeable and are grown year round, although some are more seasonal than others.

Avocados are best used raw in salads, purées and spreads — their texture and flavour are destroyed by cooking. They can also be whipped into smoothie-type drinks, and are used in cheesecakes and mousses.

Their subtle, rich smoothness is accented by a splash of lemon or lime juice, and a little salt really brings out their taste. They are lovely in salads with citrus, tomato, olives, cured meats, seafood, chicken, watercress, leafy salad greens, raw onion, chilli, coriander (cilantro), mint, basil and chives.

BANANAS

These tropical marvels have become a ubiquitous fruit-bowl item. Technically, the banana plant is the world's largest herb and is related to grasses, orchids and palms. While there are hundreds of varieties, ranging from finger-small, super-sweet bananas to large, hard, starchy, green-skinned plantains that are only good for cooking — and all the exotic white, red and purple sorts in between — the most commonly eaten type is the **cavendish**: the classic, slightly curved yellow-skinned banana.

Bananas are available year round, growing in clusters called 'hands'. They are highly nourishing, containing vitamins C and B6, iron, potassium and more digestible carbohydrate than any other fruit. In countries where they grow, the flower buds, stems and leaves are all used in cooking — in Africa, they even brew beer from bananas.

Bananas are one of the few fruits that are best ripened off the tree. If left to ripen on the plant, the skins burst and the flesh is cottony in texture. Bananas convert starch to sugar as they ripen, becoming gradually sweeter; cavendish bananas are at their eating peak when the skins are deep yellow with the occasional brown 'freckle'.

For reasons to do with transport and storage, most bananas are gas-ripened, using ethylene. Un-gassed bananas take a lot longer to ripen, but have a far richer flavour; ask your greengrocer if they can source these for you to ripen at home. Choose either just-ripe or slightly under-ripe fruit, unless you wish to use them all at once. Store them at cool room temperature — preferably away from ripe fruits, as bananas produce ethylene, which encourages ripening — but don't refrigerate as this turns their skin black and inhibits ripening. Very ripe bananas can be frozen for use in baking; peel, mash and freeze the flesh in thick, airtight freezer bags for up to 8 weeks.

Don't cut just-ripe bananas too far in advance or they will go brown (a gentle toss in lemon or lime juice will prevent this). When using bananas in pancakes, muffins, cakes and other baked goods, use over-ripe ones as these have the best taste and texture for cooking. Over-ripe bananas, which mash easily, can also be used to make ice cream, sorbet and smoothies. Firm ripe bananas are great fried in butter, finished with sugar and perhaps a slosh of brandy or rum to serve with ice cream or whipped cream.

Bananas also love brown sugar, golden syrup, honey, vanilla, yoghurt, cream, tropical fruits, coconut, cinnamon, walnuts, pecans and chocolate.

CAPSICUM (PEPPER)

A 'fruit' vegetable abundant from mid summer through early autumn, the capsicum (also called 'sweet pepper' or 'bell pepper') is from the same family as the chilli, but doesn't contain capsaicin, which gives chillies their heat. This is another vegetable for which we have Latin America to thank. Many of the world's great cuisines use this vegetable in myriad ways — raw as a salad vegetable, grilled, roasted, braised or stir-fried, made into dips, sauces, bakes and soups, or simply stuffed then baked.

Although capsicums come in colours green, red, yellow, orange and even purplish-black, they all begin life green. As they ripen, they change colour and sweeten considerably, which explains why green capsicums are the least sweet (they also contain less vitamin C). Capsicums should be heavy for their size and have smooth, glossy skins. They are prone to mould and wrinkling and should be used soon after purchase; store them at cool room temperature for no more than 2 days, or in the crisper for up to 1 week.

Capsicums are mainly peeled after roasting or grilling as the skin can be tough and chewy when cooked. They are wonderful with Mediterranean ingredients such as tomato, zucchini (courgette), eggplant (aubergine), olives, anchovies, garlic, basil, parsley, oregano, tuna, golden breadcrumbs, cheese (parmesan, pecorino, feta, goat's), red wine, vinegar (red wine and balsamic), chickpeas, cannellini beans and borlotti (cranberry) beans.

CITRUS

Where would we be without the gloriously health-giving citrus family? They are endlessly useful in cooking too, adding their unmistakable tang across the cookery spectrum in everything from cakes, muffins and breads to savoury sauces, salads, preserves and tagines and other stews.

Grapefruits

The grapefruit is usually relegated to the marmalade pot, or cut in half and eaten sprinkled with sugar for breakfast. Nowadays grapefruits are bred to be larger and less intensely sour than they once were, and in some varieties to have beautiful ruby or pink-coloured flesh. If you search, you may be able to find the characteristically smaller fruits from older varieties, which are particularly good for marmalade.

Peak grapefruit season is winter through to early spring. Choose ones that are heavy for their size and have tight, shiny skin. They are best served and stored at room temperature, and will keep for 7–10 days.

If using grapefruit in a fruit or savoury salad, use a small sharp vegetable knife to remove the bitter thick, white pith that lies under the skin, and also the membrane between each segment. The finely grated zest can stand in for lemon or orange in any citrus-based dessert.

Lemons

The lemon is a true kitchen staple. Its zest and juice are used in endless ways, either to complement other flavours or as the main attraction. The acid in lemon juice is also used to 'cook' raw fish, to tenderise meat in marinades, and is brushed over cut fruit and vegetables to stop them turning brown. Lemons are particularly high in vitamin C and cancer-fighting flavonoids, and were used in long sea voyages to prevent scurvy.

The popular **lisbon** and **eureka** lemons are oblong, with bright yellow skin and juicy, acidic, fragrant flesh. Many cooks prize the **meyer** lemon for its 'sweet' aromatic flesh and prodigious juice. Its skin and flesh are orangey-yellow and it is thought to be an orange–lemon hybrid.

When purchasing lemons, look for tight, firm, glossy skin. Avoid any that are still green as their flavour won't develop. Mould can grow quickly on lemons, spreading to other fruits, so ensure there is good air circulation. They can be stored at cool room temperature for up to 1 week, or up to 2 weeks in the refrigerator.

Many lemons have waxed skins for longer storage. Buy unwaxed ones if you can, otherwise give them a good scrub before using. If you have too many lemons to use at once, juice them and freeze the juice in ice-block trays for up to 3 months. (The juice cubes will freeze well for up to 4 months if you transfer them into a zip-lock bag once frozen.)

Lemon adds zing when used with chicken, veal, fish, olives, almonds, many vegetables and herbs, as well as rice and pasta dishes. In desserts, its flavour truly shines on its own, marries well with orange, and is set off by the richness of cream, buttery pastry and ricotta and mascarpone cheese.

Mandarins

Abundant in winter, mandarins are an old fruit native to South East Asia; they are one of the three original citrus families of which oranges, lemons, limes and grapefruits are hybrids or crosses. There are hundreds of types, but their signature feature is their thin skin, which is very easy to peel; the flesh segments also separate very easily, having minimal pith between them.

The most famous varieties are the seedless **clementine** (a cross between a mandarin and a bitter orange), the seedless and highly perfumed, slightly sweet–sour **satsuma**, from Japan (there are over 70 varieties of satsuma alone!) and the **honey murcott**, with its particularly sweet, juicy flesh.

Mandarins can dry out quickly, so choose heavy ones as they'll be juiciest. The skin should be smooth and glossy, and not too puffy. They'll keep at cool room temperature for up to 1 week and up to 12 days in the crisper.

Mandarins can be used instead of orange in any recipe, right across the cookery spectrum. The flesh cannot easily be cut away from its pith, but the fruit (for its size) yields a lot of juice. The finely grated zest can also be used.

Oranges

Oranges once appeared in children's Christmas stockings as a rare indulgence. Sadly, they are no longer considered such a special treat, perhaps because they are a fruit that generally travels well and are available year round.

Common varieties include the **navel**, so-called for its distinctive navel-shaped base; it is generally seedless and has very orange skin and very sweet, juicy flesh. The **valencia** has large, seeded fruit, and skin that tends to be green-tinged; the greening is not a sign of unripeness as oranges require cool conditions to maintain their orange colour and fruits harvested in summer may lose some of their bright hue. **Blood oranges** are becoming easier to find; their pretty red-tinged flesh yields reddish juice of an incomparable flavour — complex and even a little spicy. Blood oranges are lovely on their own, but their glorious colour can also brighten salads, jellies and sauces. The **seville** is a bitter orange, with thick skin and many seeds; it is used widely in cooking, particularly in sauces, marmalade and other preserves.

When buying oranges, avoid large bags and select fruit individually. An orange should feel heavy for its size and be firm but 'give' a little when gently squeezed. They'll keep for up to 1 week in the fruit bowl, if conditions are cool and dry, or 2 weeks in the refrigerator if kept well ventilated.

There are several ways to remove the 'zest' or thin rind of an orange (if yours has been waxed, give it a scrub first). The easiest way is to use a fine grater, taking care not to grate into the bitter white pith. If fine julienne strips are needed, you can remove wide strips of skin using a small sharp knife, working top to bottom, then lay each strip on a cutting board and carefully cut all the white pith away from the back, before finely slicing the zest.

Orange is wonderful with dried fruits, ham, turkey, game meats, duck, seafood, asparagus, beetroot (beets), sweet potato, beans, artichoke, ginger, saffron, cinnamon, cloves, caramel, chocolate, berries, honey, red wine, nuts, cream, ice cream, custard, crème fraîche and cheese (mascarpone, ricotta).

EGGPLANT (AUBERGINE)

The eggplant is native to Asia, although most of us associate it with Mediterranean and Middle Eastern cuisine, where its satisfying, meaty texture is put to excellent use. The first varieties taken to England were creamy-white, which along with their egg-like shape gave eggplant its name. Curiously, the French word *aubergine* ultimately derives from Sanskrit and means 'the vegetable that prevents flatulence'. For centuries the English only used eggplant ornamentally, believing it induced madness if consumed.

There are countless varieties, from the diminutive **pea eggplants** of South East Asia to the classic large, deep-purple **globe eggplant**. Globe eggplants should have tight, glossy skin with no blemishes, pale flesh with few visible seeds, and feel very firm and heavy. When you lightly press their flesh a dent should form, but spring back quickly. The calyx end should look healthy and green, with the calyx firmly attached. As they age, their flesh darkens and feels a little flabby, and the seeds become more pronounced.

Japanese eggplants are small, slim eggplants about 15–20 cm (6–8 inches) long and 3 cm (1¼ inches) round. Their skin is also purple, with some varieties having light mauve or speckled mauve and whitish skin. They are milder-tasting than globe eggplants and have creamy flesh lightly sprinkled with seeds. They do not require peeling and are excellent for cooking whole, for pickling, stuffing, stir-frying, grilling or steaming.

Eggplants are best stored at cool to room temperature rather than in the fridge; they'll keep for 2–3 days.

Many recipes recommend salting eggplants before cooking, to extract bitter juices. It is rare these days to come across a bitter eggplant, but salting helps improve the texture of larger eggplant. Simply scatter the sliced or diced flesh with a light dusting of salt, layer it in a large colander, set it over a sink to drain for 30 minutes, then rinse off well under cool water. Thoroughly dry the flesh with paper towels before cooking.

Eggplants go well with all the classic Mediterranean ingredients, as well as yoghurt, ricotta and North African flavours like lemon, cinnamon, chilli, coriander, cumin, paprika, almonds and walnuts.

PEARS

The pear originated in the Caucasus region millennia ago but didn't find favour until French gardeners developed new varieties in the seventeenth century, giving the creamy-fleshed, juicy characters we adore today. There are over 6000 named varieties, in all manner of shapes, colours and sizes.

Pears appear in late autumn and crop through winter. They are a very fragile fruit and easily bruised. Pears are picked when mature but still hard, then put into cold storage to 'cure'. They are brought to room temperature to ripen, and this happens from the inside out, making it difficult to know exactly how ripe they are. Once fully ripe, they make great eating for just a few days, after which their flesh turns mealy and dry.

A ripe pear should 'give' very slightly when pressed at the base of the neck; some types will exude a light, sweet fragrance. Colour is no sign of ripeness, as green varieties stay fully green when ripe. Unripe pears can be ripened at room temperature, which can take 3–10 days, or they can be refrigerated until you wish to ripen them (keep them well ventilated or they may rot). To avoid bruising, try not to stack them.

Ideal for cooking, the **beurre bosc** is an elegant, tapered, brown-skinned pear with creamy, full-flavoured flesh. The **packham** is a large green pear with white and very juicy flesh, and slightly knobbly skin that turns a little yellow when ripe. It is a good pear for eating when ripe, and cooking when a little under-ripe. The **bartlett** or **williams** pear is a good all-purpose pear similar in appearance to the packham, although it sometimes has a reddish blush when ripe. It has sweet, buttery, very juicy flesh and ripens quite quickly at room temperature. (There is a red-skinned variety too, called variously the red bartlett, red williams or red sensation.)

Pears are superb with red and dessert wines, and cheeses (blue, goat's, parmesan, sharp cheddar and ricotta), whether served raw on a cheese board, cooked with them, or tossed with them in a salad. They are also great with cloves, cinnamon, saffron, star anise, ginger, almonds, walnuts, rocket (arugula), watercress, spinach, cured meats, poultry and pork.

PINEAPPLE

The pineapple is native to South America, where its name, *ananas*, means 'fragrant, excellent fruit'. When Columbus first introduced pineapples to Europe, only the wealthy could afford them; pineapples adorned banqueting tables as a symbol of status. Later, they came to symbolise hospitality.

Pineapples are difficult to get to market ripe as they do not sweeten after harvesting and damage easily. For this reason much of the world's pineapple crop is tinned. The main distinction between the many varieties is that they are either 'smooth-leafed' or 'rough-leafed'. The rough-leafed types are smaller, with deep gold, particularly aromatic flesh, while smooth-leafed types are large and juicy, but not as sweet when ripe.

Choose one that smells sweet and fragrant. Always buy pineapples with leaves still attached and that feel heavy for their size. Store ripe pineapples at cool room temperature for 1–2 days at most, and in the refrigerator for up to 4 days in a loosely sealed plastic bag with the leaves still on.

Pineapple contains an enzyme called bromelain, which breaks down proteins and can turn meat or seafood mushy, and also inhibits gelatine setting. Cooking neutralises the enzyme, so use cooked pineapple rather than raw pineapple in gelatine-based desserts (poach it in a sugar syrup first).

Very simple methods are best for cooking pineapple — pan-frying with a little butter, sugar and alcohol (or even lightly barbecuing while brushing with a sugar–butter mixture) and serving with crème fraîche or ice cream. Of course, a perfectly ripe pineapple is wonderful served fresh for dessert.

TOMATO

Tomatoes were once a luscious height-of-summer fruit, as eagerly awaited as any seasonal treat. Unhappily, we all know what year-round supply has done to the tomato! Make it your mission to only buy ripe tomatoes in summer when they taste as they should — for the rest of the year be content with good-quality tinned tomatoes, which are excellent for cooking.

The tomato, a nightshade, is native to South America. Some rather exotic varieties have striped skin, orange-yellow or even black skin, but it is the substance lycopene (a powerful health-giving antioxidant) that gives the familiar red tomatoes their characteristic colour.

The large meaty **beefsteak (or 'garden') tomato** is at its juicy best in late summer and great in salads, sandwiches and sauces. Most common is the round, slightly squat, small to medium **salad tomato**, a good all-purpose tomato with a juicy interior and large seed cavity. The elongated, thick-skinned **roma (plum or egg) tomato** has dense, firm flesh and a not-so-juicy interior. It is the classic tomato used in Italian cookery, and for sun-drying. **Cherry tomatoes** (some round, others 'teardrop' shaped) are very small, cute and dependably sweet. They are often tossed into salads, baked onto pizza or quickly sautéed as a side dish. Finely chopped and mixed with plenty of good olive oil and chopped herbs, they also make a fine fresh 'tomato sauce' to stir through just-cooked pasta. **Green tomatoes** are end-of-season tomatoes that won't ripen fully, but are perfect for chutneys and relishes.

Tomatoes should smell fragrant and sweet — if they don't, avoid them. They should feel heavy and also be plump and shiny, with smooth, tight skins (soft spots or bruises can quickly turn mouldy). Slightly under-ripe tomatoes will fully ripen if left in a warm, sunny spot, such as a window ledge. Avoid refrigerating them as this ruins their flavour. Any surplus can always be chopped and frozen and used in cooking throughout the year.

Tomatoes are legendary with basil, garlic, mint, Italian cheeses, capsicum (pepper), balsamic vinegar, olive oil, olives, orange, eggs, bacon, chicken, beef, lamb… and of course pasta!

STUFFED VEGETABLES PROVENCAL

SERVES 6 AS A STARTER OR SIDE DISH

2 small eggplants (aubergines),
  halved lengthways
2 small zucchini (courgettes),
  halved lengthways
4 tomatoes
2 small red capsicums (peppers)
4 tablespoons olive oil
2 red onions, finely chopped
2 garlic cloves, crushed
250 g (9 oz) minced (ground) pork
250 g (9 oz) minced (ground) veal
3 tablespoons tomato paste
  (concentrated purée)
4 tablespoons white wine
2 tablespoons chopped parsley
50 g (1 3/4 oz/1/2 cup) grated parmesan
  cheese
80 g (2 3/4 oz/1 cup) fresh breadcrumbs
extra virgin olive oil, for drizzling
crusty bread, to serve

Preheat the oven to 180°C (350°F/Gas 4). Grease a large roasting tin with olive oil.

Use a spoon to hollow out the centres of the eggplants and zucchini, leaving a border around the edge. Chop the eggplant and zucchini flesh finely and set aside.

Cut the tops off the tomatoes and reserve. Use a spoon to hollow out the centres, catching the juice in a bowl, then chop the flesh roughly. Set the flesh and juice aside.

Cut the tops off the capsicums and reserve. Discard the seeds and membranes from inside the capsicum shells. Set the capsicums aside.

Heat half the olive oil in a large frying pan. Add the onion and garlic and sauté over medium–high heat for 3 minutes, or until softened. Add the pork and veal and stir for 5 minutes, or until the meat browns, breaking up any lumps with the back of a fork.

Stir in the chopped eggplant and zucchini flesh and cook for 3 minutes, then stir in the chopped tomato and reserved juice, along with the tomato paste and wine. Cook, stirring occasionally, for 10 minutes. Remove from the heat and stir in the parsley, parmesan and breadcrumbs. Season well with sea salt and freshly ground black pepper, then spoon the mixture into the eggplant, zucchini, tomato and capsicum cavities. Put the tops back on the tomatoes and capsicums.

Arrange the capsicums and eggplant in a single layer in the roasting tin. Drizzle with some of the remaining olive oil, pour 125 ml (4 fl oz/1/2 cup) water into the roasting tin and bake for 15 minutes.

Add the tomatoes to the roasting tin in a single layer and bake for 5 minutes. Finally add the zucchini in a single layer, drizzle with the remaining olive oil and bake for a further 25 minutes, or until all the vegetables are tender.

Serve hot or at room temperature, drizzled with some extra virgin olive oil, with slices of crusty bread.

Capsicum and bean stew
Serves 4–6

200 g (7 oz/1 cup) dried haricot beans
  (see Note)
2 tablespoons olive oil
1 red onion, halved and cut into thin
  wedges
2 large garlic cloves, crushed
1 red capsicum (pepper), diced
1 green capsicum (pepper), diced
2 x 400 g (14 oz) tins chopped tomatoes
2 tablespoons tomato paste (concentrated
  purée)
500 ml (17 fl oz/2 cups) vegetable stock
2 tablespoons chopped basil
125 g (4$^1/_2$ oz/$^3/_4$ cup) kalamata olives,
  pitted
1–2 teaspoons soft brown sugar
crusty bread, to serve

Put the beans in a large bowl, cover with plenty of cold water and leave to soak overnight.

Rinse the beans well, then place in a saucepan, cover with plenty of fresh cold water and bring to the boil. Reduce the heat and simmer for 45 minutes, or until just tender. Drain and set aside.

Heat the olive oil in a large heavy-based saucepan. Add the onion and garlic and sauté over medium heat for 3 minutes, or until the onion has softened a little. Add all the capsicum and cook for a further 5 minutes.

Stir in the tomatoes, tomato paste, stock and beans. Cover and simmer for 40 minutes, or until the beans are cooked through. Stir in the basil, olives and sugar, then season to taste with sea salt and freshly ground black pepper. Serve hot, with crusty bread.

NOTE: Instead of dried haricot beans you could use two 400 g (14 oz) tins of drained and rinsed haricot or borlotti (cranberry) beans. Add these at the end of cooking with the basil, olives and sugar, and just heat through.

MARINATED CAPSICUMS
SERVES 6 AS A SIDE DISH OR AS PART OF AN ANTIPASTI PLATTER

3 red capsicums (peppers)
3 thyme sprigs
1 garlic clove, thinly sliced
2 teaspoons roughly chopped flat-leaf
   (Italian) parsley
1 bay leaf
1 spring onion (scallion), sliced
1 teaspoon paprika
3 tablespoons extra virgin olive oil
2 tablespoons red wine vinegar

Preheat the oven grill (broiler) to high. Cut the capsicums into quarters and remove the seeds and membranes. Grill the capsicums, skin side up, until the skin blackens and blisters. Transfer to a bowl, cover with plastic wrap and leave until cool enough to handle. Slip the blackened skin off the capsicums, then thinly slice the flesh. Place in a bowl with the thyme, garlic, parsley, bay leaf and spring onion.

In a small bowl, whisk together the paprika, olive oil, vinegar and some sea salt and freshly ground black pepper. Pour over the capsicum mixture and toss to combine well. Cover and refrigerate for at least 3 hours, or preferably overnight, to allow the flavours to develop.

Serve at room temperature as part of an antipasti selection, or as an accompaniment to grilled meats or fish. Marinated capsicums will keep, covered well and refrigerated, for 3 days.

ALGERIAN EGGPLANT JAM
SERVES 6–8 AS A STARTER OR SIDE DISH

2 eggplants (aubergines), about
   400 g/14 oz in total
olive oil, for pan-frying
2 garlic cloves, crushed
1 teaspoon sweet paprika
1½ teaspoons ground cumin
½ teaspoon sugar
1 tablespoon lemon juice

Cut the eggplants into 1 cm (½ inch) slices. Layer the slices in a colander in the sink, sprinkling each layer lightly with sea salt. Leave to stand for 30 minutes. Rinse well, then dry thoroughly using paper towels.

Heat 5 mm (¼ inch) of olive oil in a large frying pan. Add the eggplant in batches and cook over medium heat until golden brown on both sides. Drain well on paper towels, then chop finely. Place in a colander until any remaining oil has drained off, then transfer to a bowl and add the garlic, paprika, cumin and sugar.

Wipe the pan clean, then add the eggplant mixture and stir constantly over medium heat for 2 minutes. Transfer to a serving bowl, stir in the lemon juice and season to taste with sea salt and freshly ground black pepper.

Serve at room temperature with roast lamb or chicken. Algerian eggplant jam can be refrigerated for 3–4 days.

MOROCCAN EGGPLANT WITH COUSCOUS

SERVES 4 AS A SIDE DISH

185 g (6½ oz/1 cup) instant couscous
200 ml (7 fl oz) olive oil
1 onion, halved and sliced
1 eggplant (aubergine)
3 teaspoons ground cumin
1½ teaspoons garlic salt
¼ teaspoon ground cinnamon
1 teaspoon paprika
¼ teaspoon ground cloves
½ teaspoon sea salt
50 g (1¾ oz) unsalted butter
a handful of roughly chopped parsley

Put the couscous in a large bowl and add 375 ml (13 fl oz/ 1½ cups) boiling water. Cover and leave for 10 minutes, then fluff up the grains with a fork.

Heat 2 tablespoons of the olive oil in a large frying pan. Add the onion and sauté over medium heat for 8–10 minutes, or until browned. Remove from the pan, reserving the pan.

Cut the eggplant into 1 cm (½ inch) slices, then cut the slices into quarters. Place in a large bowl. In a small bowl mix together the cumin, garlic salt, cinnamon, paprika, cloves and sea salt, then sprinkle over the eggplant, tossing to coat well.

Add the remaining oil to the pan and reheat the pan over medium heat. Add the eggplant and cook, turning once, for 20 minutes, or until soft and browned. Remove from the pan and set aside to cool.

Melt the butter in the same frying pan. Add the couscous and cook, stirring, for 2–3 minutes, then stir in the onion, eggplant and parsley. Serve at room temperature.

EGGPLANT PARMIGIANA

SERVES 6–8 AS A SIDE DISH OR LIGHT MEAL

1.25 kg (2 lb 12 oz) tomatoes

olive oil, for pan-frying

1 onion, diced

2 garlic cloves, crushed

1 kg (2 lb 4 oz) eggplants (aubergines), thinly sliced

250 g (9 oz) bocconcini (fresh baby mozzarella cheese), sliced

185 g (6 oz/1½ cups) finely grated cheddar cheese

a large handful of basil leaves, torn

50 g (1¾ oz/½ cup) grated parmesan cheese

Bring a saucepan of water to the boil. Using a small sharp knife, score a small cross in the base of each tomato. Place the tomatoes in the boiling water for about 20 seconds, remove using a slotted spoon, then plunge into a bowl of iced water. Drain the tomatoes and peel the skins away from the cross. Cut the tomatoes in half, scoop out the seeds with a teaspoon and roughly chop the flesh.

Heat 3 tablespoons of olive oil in a large saucepan. Add the onion and sauté over medium heat for 5 minutes, or until softened. Add the garlic and cook for 1 minute, then add the tomato and simmer for 15 minutes. Season to taste with sea salt.

Meanwhile, preheat the oven to 200°C (400°F/Gas 6).

Place a large frying pan over medium heat and pour in enough olive oil to cover the base. Fry the eggplant in batches for 3–4 minutes, or until golden brown, adding more oil to the pan as needed. Drain on paper towels.

Place one-third of the eggplant in a 1.75 litre (61 fl oz/ 7 cup) baking dish. Top with half the bocconcini and cheddar. Repeat the layers, finishing with a layer of eggplant. Pour the tomato mixture over the top, then scatter with the basil and parmesan. Bake for 40 minutes, or until the eggplant is tender. Serve hot.

VARIATION: If you would prefer not to fry the eggplant in oil, brush it lightly with olive oil and brown lightly under a hot oven grill (broiler).

CAPONATA

SERVES 6 AS A SIDE DISH OR AS PART OF AN ANTIPASTI PLATTER

1 kg (2 lb 4 oz) tomatoes
3 tablespoons olive oil
2 onions, sliced
2 red capsicums (peppers), thinly sliced
4 garlic cloves, finely chopped
4 celery stalks, sliced
1 large eggplant (about 500 g/1 lb 2 oz),
   diced
2 tablespoons thyme leaves
2 tablespoons caster (superfine) sugar
125 ml (4 fl oz/$1/2$ cup) red wine vinegar
125 g ($4^{1}/_{2}$ oz/$1/2$ cup) pitted green olives,
   rinsed well and drained
2 tablespoons capers, rinsed and drained

Bring a saucepan of water to the boil. Using a small sharp knife, score a small cross in the base of each tomato. Place the tomatoes in the boiling water for about 20 seconds, remove using a slotted spoon, then plunge into a bowl of iced water. Drain the tomatoes and peel the skins away from the cross. Cut the tomatoes in half, scoop out the seeds with a teaspoon and roughly chop the flesh.

Heat the olive oil in a large heavy-based frying pan. Add the onion, capsicum, garlic, celery and eggplant. Cover and cook over low heat for 20–30 minutes, or until tender, stirring occasionally. Season to taste with sea salt and freshly ground black pepper.

Remove the lid, add the tomato and thyme and simmer, uncovered, for a further 15 minutes.

Stir in the sugar, vinegar, olives and capers and mix well. Taste and season again if needed.

Serve warm or at room temperature as part of an antipasti selection or as an accompaniment to grilled meats. Caponata is also delicious stirred through drained, cooked pasta and served topped with grated pecorino cheese as a light lunch or dinner.

Slow-roasted balsamic tomatoes

Makes 40 pieces

10 firm roma (plum) tomatoes
8 garlic cloves, crushed
4 tablespoons caster (superfine) sugar
4 tablespoons torn basil leaves
1 tablespoon chopped oregano leaves
a few drops of good-quality balsamic
  vinegar

Preheat the oven to 140°C (275°F/Gas 1). Line two baking trays with baking paper. Cut each tomato lengthways into quarters and arrange in rows on the baking trays.

In a small bowl, mix together the garlic, sugar, basil, oregano and vinegar. Using your fingers, press a little of the mixture onto the sides of each tomato quarter and season with sea salt and freshly ground black pepper.

Bake for 2½ hours, or until the tomatoes are slightly shrivelled around the edges and semi-dried, but still soft in the middle.

Serve warm or at room temperature as part of an antipasti selection, or with barbecued meats.

Slow-roasted balsamic tomatoes will keep in an airtight container in the refrigerator for up to 1 week.

TOMATO BREAD SOUP
SERVES 4

750 g (1 lb 10 oz) vine-ripened tomatoes
1 loaf (about 450 g/1 lb) of day-old,
  good-quality rustic bread (see Note)
1 tablespoon olive oil
3 garlic cloves, crushed
1 tablespoon tomato paste (concentrated
  purée)
1.25 litres (44 fl oz/5 cups) vegetable
  stock or water
4 tablespoons torn basil leaves
2–3 tablespoons extra virgin olive oil,
  plus extra, for drizzling

Bring a saucepan of water to the boil. Using a small sharp knife, score a small cross in the base of each tomato. Place the tomatoes in the boiling water for about 20 seconds, remove using a slotted spoon, then plunge into a bowl of iced water. Drain the tomatoes and peel the skins away from the cross. Cut the tomatoes in half, scoop out the seeds with a teaspoon, then roughly chop the flesh.

Trim the crust from the bread, then tear the bread into 3 cm (1¼ inch) chunks.

Heat the olive oil in a large saucepan. Add the garlic, tomato and tomato paste, then simmer, stirring occasionally, for 10–15 minutes, or until the mixture has reduced.

Add the stock and bring to the boil, stirring for 2–3 minutes. Reduce the heat to medium, add the bread and cook, stirring often, for 5 minutes, or until the bread softens and absorbs most of the liquid. Add more stock or water if the soup is too thick.

Remove from the heat, stir in the basil and extra virgin olive oil, and leave to stand for 5 minutes for the flavours to develop. Serve drizzled with a little extra virgin olive oil.

NOTE: It is important to use a quality loaf from a good baker, as mass-produced breads do not go slightly stale after 1 day. Their texture is generally too soft and smooth to work well in this recipe.

Rice-stuffed tomatoes

Makes 8

8 tomatoes
110 g (3³/₄ oz/¹/₂ cup) short-grain
  white rice
2 tablespoons olive oil, plus extra,
  for brushing
1 red onion, finely chopped
1 garlic clove, crushed
1 teaspoon dried oregano
3 tablespoons pine nuts
3 tablespoons currants
a handful of chopped basil
2 tablespoons chopped flat-leaf (Italian)
  parsley
1 tablespoon chopped dill

Slice the top off each tomato and reserve. Scoop out the flesh using a teaspoon and place in a strainer over a bowl to drain the juice. Reserve the juice. Finely chop the flesh and place in a bowl. Stand the tomato shells upside down on a rack to drain.

Cook the rice in a saucepan of lightly salted boiling water for 10–12 minutes, or until just tender. Drain well, then set aside to cool.

Meanwhile, preheat the oven to 160°C (315°F/Gas 2–3). Lightly oil a large baking dish.

Heat the olive oil in a frying pan. Add the onion, garlic and oregano and sauté over medium heat for 8 minutes, or until the onion has softened. Add the pine nuts and currants and sauté for a further 5 minutes. Remove from the heat and stir in the basil, parsley and dill. Season to taste with sea salt and freshly ground black pepper.

Stir the onion mixture and reserved tomato flesh into the cooled rice. Fill the tomato shells with the rice mixture, mounding it slightly at the top. Spoon 1 tablespoon of the reserved tomato juice on top of each tomato and replace the tomato tops.

Lightly brush the tomatoes with olive oil and arrange in the baking dish. Bake for 20–30 minutes, or until cooked and heated through. Serve warm or at room temperature.

Avocado and grapefruit salad
SERVES 4

2 ruby grapefruits
1 ripe avocado
200 g (7 oz) watercress sprigs
1 French shallot, finely sliced
1 tablespoon sherry vinegar
3 tablespoons olive oil

Using a small sharp knife, peel each grapefruit, taking care to remove all the white pith. Working over a bowl to catch any juices for the dressing, carefully remove the grapefruit segments by cutting between the white membrane and the flesh. Squeeze out any juice remaining in the membranes into the bowl.

Peel the avocado, cut it in half and remove the stone. Cut the flesh into 2 cm (3/4 inch) wedges and place in a bowl with the grapefruit segments, watercress and shallot.

Put 1 tablespoon of the reserved grapefruit juice in a small bowl with the vinegar, olive oil and a little sea salt and freshly ground black pepper. Whisk together well, then pour the dressing over the salad and toss gently to coat.

Divide among serving plates and serve immediately.

Avocado with lime and chilli
SERVES 6 AS A STARTER OR SIDE DISH

2 ripe avocados
1 teaspoon finely grated lime or
   lemon zest
2 tablespoons lime or lemon juice
1 teaspoon soft brown sugar
1 tablespoon olive oil
1 tablespoon chopped flat-leaf (Italian)
   parsley
2–3 small red chillies, seeded and diced

Just before serving, peel the avocados, then cut them in half and remove the stones. Cut into slices or wedges and arrange on serving plates.

Put the remaining ingredients in a small bowl. Add a little sea salt and freshly ground black pepper and whisk until well combined. Pour the dressing over the avocado and serve immediately.

Avocado and grapefruit salad

Parmesan pears
SERVES 6

3 ripe firm pears, such as packham or
  beurre bosc
40 g (1¹/₂ oz) unsalted butter
6 thin slices of pancetta, finely chopped
2 spring onions (scallions), finely sliced
60 g (2¹/₄ oz/³/₄ cup) fresh white
  breadcrumbs
4 tablespoons grated parmesan cheese

Heat the oven grill (broiler) to medium–high. Cut the pears in half lengthways and remove the cores.

Melt the butter in a frying pan. Brush the pears with a little of the melted butter and place, cut side up, on a baking tray or grill tray. Grill for 4 minutes, or until starting to brown on top.

Add the pancetta and spring onion to the remaining butter in the pan. Sauté over medium heat for 3 minutes, or until the spring onion is soft but not brown. Stir in the breadcrumbs and some freshly ground black pepper to taste.

Spoon the pancetta mixture into the pear cavities, sprinkle with the parmesan and grill for 3 minutes, or until the cheese is golden brown.

Serve warm as a starter, or as an accompaniment to roast chicken.

Rocket, grape and walnut salad
SERVES 6

1 butter lettuce
1 radicchio
155 g (5¹/₂ oz) rocket (arugula) leaves
180 g (6 oz/1 cup) green seedless grapes
60 g (2¹/₄ oz/¹/₂ cup) broken walnuts,
  toasted

DRESSING
4 tablespoons extra virgin olive oil
1 tablespoon lemon juice
2 teaspoons wholegrain mustard
1 tablespoon snipped chives

Discard the tough outer leaves from the lettuce and radicchio, then separate the remaining leaves. Gently wash the lettuce, radicchio and rocket, then dry thoroughly. Transfer to an airtight container or sealed plastic bag and chill.

Put the chilled leaves in a large bowl with the grapes. Toss well, then scatter with the walnuts.

To make the dressing, put the olive oil, lemon juice and mustard in a bowl and whisk together well. Season with freshly ground black pepper and stir in the chives.

Drizzle the dressing over the salad and serve.

Scallop, ginger and spinach salad
Serves 4

300 g (10½ oz) scallops
100 g (3½ oz/2 cups) baby English
  spinach leaves
1 small red capsicum (pepper), cut into
  very fine strips
50 g (1¾ oz/½ cup) bean sprouts, tails
  trimmed
olive oil, for brushing

DRESSING
25 ml (1 fl oz) sake or dry sherry
1 tablespoon lime or lemon juice
2 teaspoons shaved palm sugar (jaggery)
  or soft brown sugar
1 teaspoon fish sauce

Using a small sharp knife, carefully remove any membrane from the scallops. Pat the scallops dry with paper towels.

Put the dressing ingredients in a small bowl and mix together well. Set aside.

Divide the spinach, capsicum and bean sprouts among serving plates.

Heat a chargrill pan or barbecue hotplate to medium, then lightly brush with olive oil. Cook the scallops in batches for 1 minute on each side, or until just cooked through.

Arrange the scallops over the salad, drizzle with the dressing and serve.

LEMON, HERB AND FISH RISOTTO
SERVES 4

1.25 litres (44 fl oz/5 cups) fish stock
60 g (2¼ oz) unsalted butter
400 g (14 oz) skinless firm white fish
   fillets, cut into 3 cm (1¼ inch) cubes
1 onion, finely chopped
1 garlic clove, crushed
a large pinch of saffron threads
330 g (11¾ oz/1½ cups) risotto rice
2 tablespoons lemon juice
1 tablespoon chopped flat-leaf (Italian)
   parsley
1 tablespoon snipped chives
1 tablespoon chopped dill
lemon slices, to garnish
herb sprigs, to garnish

Pour the stock into a saucepan and bring to the boil.
Reduce the heat, then cover and keep at simmering point.

Melt half the butter in a frying pan. Add the fish in
batches and fry over medium–high heat for 3–4 minutes, or
until the fish is just cooked through, turning once. Remove
from the pan and set aside. Keep warm.

Melt the remaining butter in a large heavy-based
saucepan. Add the onion and garlic and sauté over medium
heat for 5 minutes, or until the onion has softened. Add
the saffron and rice and stir to coat, then add 125 ml
(4 fl oz/½ cup) of the simmering stock and cook, stirring
constantly, over low heat until all the stock has been absorbed.
Continue adding the stock, 125 ml (4 fl oz/½ cup) at a
time, stirring constantly and making sure the stock has been
absorbed before adding more. Cook for 20–25 minutes, or
until the rice is tender and creamy; you may need slightly less
or more stock.

Stir in the lemon juice, parsley, chives and dill. Add the
fish and stir gently. Spoon into warmed serving bowls, garnish
with lemon slices and herb sprigs and serve.

PRESERVED LEMONS
FILLS A 2 LITRE (70 FL OZ/8 CUP) JAR

8–12 small thin-skinned lemons
315 g (11 oz/1 cup) rock salt
750 ml (26 fl oz/3 cups) lemon juice
   (10–12 lemons should yield this
   amount of juice)
1/2 teaspoon black peppercorns
1 bay leaf
olive oil, for covering

Scrub the lemons under warm running water with a soft brush to remove the wax coating if necessary.

Starting from the top and cutting almost to the base, cut the lemons into quarters, taking care not to cut all the way through. Gently open each lemon, remove any visible seeds and pack 1 tablespoon of the rock salt inside each lemon. Push the lemons back into shape and pack tightly into a 2 litre (70 fl oz/8 cup) sterilised jar (see Note below) with a tight-fitting lid. The lemons should be firmly packed and fill the jar (depending on their size, you may not need all 12).

Add 250 ml (9 fl oz/1 cup) of the lemon juice, the peppercorns, bay leaf and remaining rock salt to the jar. Fill the jar to the top with the remaining lemon juice. Seal and leave in a cool, dark place for 6 weeks, inverting the jar each week to dissolve the salt. The liquid will be cloudy initially, but will clear by the fourth week.

To test if the lemons are preserved, cut through the centre of one of the lemon quarters. If the pith is still white, the lemons aren't quite ready. In this case, re-seal and leave for another week before testing again. The lemons should be soft-skinned and the pith translucent.

Once the lemons are preserved, cover the brine with a layer of olive oil. Replace the oil each time you remove some of the lemon pieces so that the lemons remain covered with oil. Refrigerate after opening.

NOTE: Preserved lemons can be stored for up to 6 months in a cool, dark place. Only the rind is used in cooking. Discard the salty flesh and bitter pith, then rinse and finely slice or chop the rind and use to flavour couscous, stuffings, tagines and casseroles.

Jars must always be sterilised before pickles, preserves or jams are put in them for storage, otherwise bacteria will multiply. To sterilise your jars and lids, rinse them with boiling water and place in a warm oven for 20 minutes, or until completely dry. (Jars with rubber seals are safe to warm in the oven and won't melt.) Never dry your jars with a tea towel (dish towel) — even a clean one may have germs on it and contaminate the jars.

PEAR TARTE TATIN
SERVES 8

145 g (5 oz/²/₃ cup) caster (superfine)
   sugar
50 g (1³/₄ oz) unsalted butter, chopped
¹/₂ teaspoon ground ginger
¹/₂ teaspoon ground cinnamon
3 beurre bosc pears, peeled, cored and cut
   into sixths, widthways
450 g (1 lb) block of frozen puff pastry,
   thawed
thick (double/heavy) cream, to serve

Preheat the oven to 220°C (425°F/Gas 7).

Place a 22 cm (8¹/₂ inch) heavy-based frying pan with an ovenproof handle over medium heat. Add the sugar and heat, shaking the pan constantly, until the sugar is a dark caramel colour. Stir in the butter, ginger and cinnamon. Arrange the pears on top, spoon the syrup over to coat, then reduce the heat to low and cook, covered, for 5 minutes, or until the pears just begin to soften.

Remove the frying pan from the heat and rearrange the pears over the base of the pan, overlapping so they fit tightly and look neat. Leave to cool slightly.

On a lightly floured work surface, roll out the pastry to a 24 cm (9¹/₂ inch) round. Place the pastry over the pears in the frying pan, tucking the edges down the side of the pan to enclose the pears.

Transfer to the oven and bake for 20–25 minutes, or until the pastry is golden and puffed.

Remove from the oven and leave to stand in the pan for 10 minutes, then run a knife around the edge of the pan to loosen the tart and invert it onto a serving platter.

Serve warm, with cream.

Pear tarte tatin is best served the day it is made.

VARIATION: To make an apple tarte tatin, replace the pears with 2–3 green apples (such as granny smiths).

MACERATED ORANGES
SERVES 4

4 oranges
1 teaspoon grated lemon zest
1 tablespoon lemon juice
3 tablespoons caster (superfine) sugar
2 tablespoons Cointreau or maraschino
   (cherry liqueur), optional

Cut a thin slice from both ends of the oranges. Using a small sharp knife, peel each orange, taking care to remove all the white pith. Working over a bowl to catch any juices for the dressing, carefully remove the orange segments by cutting between the white membrane and the flesh. Squeeze out any juice remaining in the membranes into the bowl.

Put the orange segments and reserved juice in a serving dish or bowl and sprinkle with the lemon zest, lemon juice and sugar. Stir gently to combine well. Cover and refrigerate for at least 2 hours.

Stir the mixture again. Just before serving, stir in the liqueur, if desired. Serve chilled.

MANDARIN ICE
SERVES 4–6

10 mandarins
115 g (4 oz/$\frac{1}{2}$ cup) caster (superfine)
   sugar

Squeeze the mandarins to make 500 ml (17 fl oz/2 cups) juice. Strain the juice and set aside.

Put the sugar and 250 ml (9 fl oz/1 cup) water in a small saucepan. Stir over low heat until the sugar has dissolved, then simmer for 5 minutes. Remove from the heat and leave to cool slightly.

Stir the mandarin juice into the sugar syrup, then pour into a shallow metal tray. Freeze for 2 hours, or until frozen.

Transfer to a food processor and blend until slushy, or transfer to a bowl and beat with a wooden spoon.

Return to the freezer and repeat this process three more times. Serve in chilled glasses.

Mandarin ice is best eaten within 1–2 days of making.

ORANGE AND LEMON SYRUP CAKE

SERVES 10–12

3 lemons

3 oranges

250 g (9 oz) cold unsalted butter, chopped

690 g (1 lb 8$^1/_2$ oz/3 cups) caster (superfine) sugar

6 eggs, lightly beaten

375 ml (13 fl oz/1$^1/_2$ cups) milk

375 g (13 oz/3 cups) self-raising flour, sifted

Preheat the oven to 160°C (315°F/Gas 2–3). Grease a 24 cm (9$^1/_2$ inch) spring-form cake tin and line the base and side with baking paper.

Finely grate the zest from the lemons and oranges to give 3 tablespoons of each, then squeeze the fruit to give 185 ml (6 fl oz/$^3/_4$ cup) lemon juice and 185 ml (6 fl oz/$^3/_4$ cup) orange juice.

Put the butter, 500 g (1 lb 2 oz/2 cups) of the sugar and 1 tablespoon each of the lemon and orange zest in a saucepan over low heat. Stir until the butter has melted and the sugar has dissolved, then pour into a bowl.

Add half the beaten egg, half the milk and half the flour, beating with electric beaters until just combined. Add the remaining egg, milk and flour and beat until just smooth — do not overmix.

Pour the batter into the prepared cake tin and bake for 1$^1/_4$ hours, or until a cake tester inserted into the centre of the cake comes out clean — cover the cake with foil if it browns too quickly. Remove the cake from the oven and allow to cool in the tin.

Put all the fruit juice in a saucepan with the remaining citrus zest, remaining sugar and 125 ml (4 fl oz/$^1/_2$ cup) water. Stir over low heat until the sugar has dissolved, then increase the heat and boil for 10 minutes, or until the mixture reduces and thickens slightly.

Pour the hot syrup over the cooled cake. Leave in the tin for a further 10 minutes, then invert onto a serving plate.

Orange and lemon syrup cake will keep for 4 days, stored in a cool place in an airtight container.

CHILLED MANDARIN SOUFFLE

SERVES 4

vegetable oil, for brushing
5 eggs, separated
230 g (8 oz/1 cup) caster (superfine) sugar
2 teaspoons finely grated mandarin zest
185 ml (6 fl oz/³/4 cup) strained mandarin juice
1 tablespoon powdered gelatine
310 ml (10³/4 fl oz/1¹/4 cups) pouring (whipping) cream, lightly whipped, plus extra, for serving
julienned mandarin zest, to garnish

Cut out four wide strips of foil, then fold each in half lengthways. Wrap the foil strips around the outside of four 250 ml (9 fl oz/1 cup) soufflé dishes, positioning the foil so it extends 4 cm (1¹/2 inches) above the rims. Secure with string. Brush the inside of the foil with a little oil.

In a small bowl, beat the egg yolks, sugar and mandarin zest using electric beaters for 3 minutes, or until the sugar has dissolved and the mixture is thick and pale.

Heat the mandarin juice in a small saucepan. Whisking continuously, gradually add it to the egg yolk mixture until well combined.

Sprinkle the gelatine over 3 tablespoons water in a small heatproof bowl, then leave to stand for 5 minutes, or until the gelatine is soft. Stand the bowl in a small saucepan of barely simmering water and heat for 3 minutes, or until the gelatine has dissolved. Gradually add the gelatine to the mandarin mixture, whisking gently until combined.

Transfer to a large bowl, cover with plastic wrap and refrigerate for 15 minutes, or until the mixture has thickened but has not set. Using a metal spoon, fold the whipped cream into the mandarin mixture until almost combined.

Using electric beaters, whisk the egg whites in a clean, dry bowl until soft peaks form. Fold the beaten egg white quickly and lightly into the mandarin mixture until the mixture is just combined, with no streaks of egg white remaining.

Gently spoon into the prepared soufflé dishes and refrigerate for 4 hours, or until set.

To serve, remove the foil collar, then decorate with whipped cream and julienned mandarin zest.

NOTE: This soufflé can be made up to 8 hours ahead of serving.

RICOTTA CREPES WITH ORANGE SAUCE
SERVES 4

85 g (3 oz/²/₃ cup) plain (all-purpose)
   flour
1 egg, lightly beaten
330 ml (11¼ fl oz/1¹/₃ cups) milk
butter, for greasing

FILLING
3 tablespoons sultanas (golden raisins)
250 ml (9 fl oz/1 cup) orange juice
200 g (7 oz/heaped ³/₄ cup) ricotta cheese
1 teaspoon finely grated orange zest
¼ teaspoon natural vanilla extract

ORANGE SAUCE
50 g (1³/₄ oz) unsalted butter
3 tablespoons caster (superfine) sugar
1 tablespoon Grand Marnier

Sift the flour and a pinch of sea salt into a bowl and make a
well in the centre. In another bowl combine the egg and milk,
then add to the well in the flour and beat using a wire whisk
until a smooth batter forms. Cover with plastic wrap and leave
to stand for 30 minutes.

Heat a 16 cm (6¼ inch) crepe or non-stick frying pan.
Lightly grease with butter, then pour 2–3 tablespoons of
batter into the pan, swirling the pan so the mixture coats
the base evenly. Cook over medium heat for 1–2 minutes,
or until golden underneath. Turn and cook the other side for
30 seconds, then transfer to a plate. Repeat with the remaining
batter to make eight crepes, greasing the pan lightly as needed
and stacking the crepes on a plate with a sheet of baking paper
between each.

Preheat the oven to 160°C (315°F/Gas 2–3).

To make the filling, put the sultanas in a small bowl, pour
the orange juice over and leave to soak for 15 minutes. Drain
the sultanas well, reserving the juice, and place in a larger bowl
with the ricotta, orange zest and vanilla. Mix together well.

Place large tablespoons of the filling at the edge of each
crepe. Fold in half, then in half again. Divide the crepes among
four ovenproof serving plates and bake for 10 minutes.

Meanwhile, make the orange sauce. Melt the butter in
a small saucepan over low heat. Add the sugar and reserved
orange juice and stir over medium heat without boiling until
the sugar has dissolved. Bring to the boil, then reduce the heat
and simmer for 10 minutes, or until reduced slightly. Stir
in the Grand Marnier and allow to cool for 3–4 minutes.

Pour the orange sauce over the warm crepes and
serve immediately.

NOTE: The crepes may be cooked up to 4 hours in advance. Cover
and refrigerate until required, then fill and heat just before serving.

THREE-FRUIT MARMALADE
FILLS TWELVE 250 ML (9 FL OZ/1 CUP) JARS

1 grapefruit
2 oranges
2 lemons
3 kg (6 lb 12 oz) sugar

Scrub the grapefruit, oranges and lemons under warm running water with a soft brush to remove the wax coating if necessary.

Cut the grapefruit into quarters, and the oranges and lemons in half. Slice the fruit very thinly, reserving any seeds, and place the slices in a large, non-metallic bowl. Put the seeds in a piece of muslin (cheesecloth), then tie securely with string. Add to the bowl with 2.5 litres (87 fl oz/10 cups) water, then cover and leave overnight.

Put two small plates in the freezer. Transfer the fruit and water to a large heavy-based saucepan or preserving pan. Bring slowly to the boil, then reduce the heat, cover and simmer for 1 hour, or until the fruit is tender. Meanwhile, warm the sugar slightly by spreading it in a large baking dish and heating in a 120°C (235°F/Gas ½) oven for 10 minutes, stirring occasionally.

Add the warmed sugar to the fruit all at once. Stir over low heat, without boiling, for 5 minutes, or until the sugar has dissolved. Bring the mixture to the boil, then allow to boil rapidly for 50–60 minutes. Start testing for setting point by placing a little of the hot marmalade on a chilled plate. When setting point is reached, a skin will form on the surface and will leave a clear trail when you push it with your finger. If the marmalade doesn't set, keep cooking and testing until it does. Leave to cool for 10 minutes, then skim off any impurities that have risen to the surface. Remove the muslin bag.

Spoon the slightly cooled marmalade into hot, sterilised jars (see Note below) and seal. Turn the jars upside down for 2 minutes, then turn back up again and leave to cool completely. Label and date for storage.

Store in a cool, dark place for 12 months. Once opened, three-fruit marmalade will keep in the refrigerator for 8 weeks.

NOTE: Jars must always be sterilised before pickles, preserves or jams are put in them for storage, otherwise bacteria will multiply. To sterilise your jars and lids, rinse them with boiling water and place in a warm oven for 20 minutes, or until completely dry. (Jars with rubber seals are safe to warm in the oven and won't melt.) Never dry your jars with a tea towel (dish towel) — even a clean one may have germs on it and contaminate the jars.

CARAMELISED APPLE MOUSSE
SERVES 4

50 g (1³/₄ oz) unsalted butter
3 tablespoons caster (superfine) sugar
170 ml (5¹/₂ fl oz/²/₃ cup) pouring
  (whipping) cream
500 g (1 lb 2 oz) green apples, peeled,
  cored and cut into thin wedges
2 eggs, separated

Put the butter and sugar in a heavy-based frying pan and stir over low heat until the sugar has dissolved. Increase the heat to medium and cook, stirring frequently, until the mixture turns deep golden. Add 2 tablespoons of the cream and stir until smooth.

Add the apple wedges and cook over medium heat, stirring frequently, for 10–15 minutes, or until they are tender and the caramel is very reduced and sticky. Remove eight apple wedges and set aside as a garnish.

Transfer the remaining apples and caramel to a food processor and blend until smooth. Tip into a large bowl, then stir in the egg yolks and leave to cool.

Using electric beaters, whisk the egg whites in a clean, dry bowl until soft peaks form, then fold the egg white into the cooled apple mixture.

Whip the remaining cream until firm peaks form, then fold into the apple mixture. Pour into a 750 ml (26 fl oz/3 cup) serving bowl or four 185 ml (6 fl oz/³/₄ cup) individual serving moulds. Refrigerate for 3 hours, or until firm.

Serve garnished with the reserved apple wedges.

APPLE AND PEAR SORBET
SERVES 4–6

4 large green apples, peeled, cored
  and chopped
4 pears, peeled, cored and chopped
1 long, thick strip of lemon zest
1 cinnamon stick
3 tablespoons lemon juice
4 tablespoons caster (superfine) sugar
2 tablespoons Calvados or Poire William
  liqueur (optional)

Put the apple and pear in a large saucepan with the lemon zest, cinnamon stick and enough water to just cover the fruit. Bring to a simmer, cover, then cook over low–medium heat for 6–8 minutes, or until the fruit is tender. Strain, reserving 4 tablespoons of the cooking liquid, and discard the lemon zest and cinnamon stick.

Transfer the fruit to a food processor, add the lemon juice and blend until smooth.

Put the sugar and reserved cooking liquid in a saucepan, bring to the boil and simmer for 1 minute. Stir in the fruit purée and the liqueur, if using.

Pour into a shallow metal tray and freeze for 2 hours, or until the mixture is frozen around the edges.

Transfer to a food processor and blend until just smooth, or transfer to a bowl and beat with a wooden spoon until smooth. Return to the tray and freeze for a further 2 hours, then process or beat again.

Repeat this process two more times.

Transfer the sorbet to an airtight container, cover the surface with a piece of baking paper, then seal with a lid. Freeze until firm.

Serve in small glasses or bowls.

Apple and pear sorbet can be frozen for up to 3 days.

NOTE: You can pour an extra nip of Calvados over the sorbet to serve, if desired.

SWEET DRUNKEN PINEAPPLE
SERVES 6

1 large pineapple
oil, for brushing
40 g (1¹/₂ oz/¹/₄ cup) coarsely grated palm
   sugar (jaggery) or soft brown sugar
2¹/₂ tablespoons rum
2 tablespoons lime juice
3 tablespoons small mint leaves
thick (double/heavy) cream, to serve

Preheat a barbecue grill plate or chargrill pan to medium.

Trim and peel the pineapple, remove the 'eyes', then cut the pineapple into quarters lengthways.

Brush the hot grill plate or pan with oil, add the pineapple quarters and cook for about 10 minutes, turning to brown all over.

Remove the pineapple from the heat and cut each piece into slices 1.5 cm (⁵/₈ inch) thick. Overlap the slices on a large serving plate.

Put the sugar, rum and lime juice in a small pouring jug, stirring to dissolve the sugar. Pour the mixture evenly over the warm pineapple slices, then cover with plastic wrap and refrigerate for several hours.

Serve at room temperature, sprinkled with the mint leaves and with some cream passed separately.

SPICED BAKED APPLES
SERVES 4

melted butter, for brushing
4 green apples
3 tablespoons raw (demerara) sugar
3 tablespoons chopped dried figs
3 tablespoons chopped dried apricots
3 tablespoons slivered almonds
1 tablespoon apricot jam
¼ teaspoon ground cardamom
¼ teaspoon ground cinnamon
30 g (1 oz) unsalted butter, chopped
whipped cream, custard or ice cream,
   to serve (optional)

Preheat the oven to 180°C (350°F/Gas 4). Brush a square, deep baking dish with melted butter.

Peel the apples and remove the cores. Gently roll each apple in the sugar. In a bowl, mix together the figs, apricots, almonds, jam and spices.

Fill each apple with some of the fruit mixture. Place the apples in the baking dish and dot with pieces of butter.

Bake for 35–40 minutes, or until the apples are tender. Serve warm with whipped cream, custard or ice cream, if desired.

Spiced baked apples are best prepared and baked just before serving.

BANANA FRITTERS IN COCONUT BATTER

SERVES 6

100 g (3½ oz/½ cup) glutinous rice flour
  (see Note)
100 g (3½ oz/1 cup) freshly grated
  coconut, or 60 g (2¼ oz/⅔ cup)
  desiccated coconut
3 tablespoons sugar
1 tablespoon sesame seeds
3 tablespoons coconut milk
oil, for deep-frying
3 firm, ripe bananas
ice cream, to serve

Put the rice flour, coconut, sugar, sesame seeds, coconut milk and 3 tablespoons water in a bowl and whisk until a smooth batter forms, adding a little more water if the batter is too thick. The batter should have a thick, coating consistency. Cover with plastic wrap and leave to stand for 1 hour.

Fill a wok or deep heavy-based saucepan one-third full of oil and heat to 180°C (350°F), or until a cube of bread dropped into the oil browns in 15 seconds.

Peel the bananas and cut them in half lengthways, then cut in half crossways. Working in batches, dip each piece of banana into the batter, allowing the excess batter to drain off, then gently drop into the hot oil. Cook for 4–6 minutes, or until golden brown all over. Remove the fritters with a slotted spoon and drain well on paper towels. Serve hot with ice cream.

NOTE: Glutinous rice flour is available from Asian food stores and is made by finely grinding glutinous or sticky rice; do not confuse it with the flour made from ordinary rice.

CHOCOLATE BANANA CAKE
SERVES 6–8

3 very ripe bananas, mashed
170 g (6 oz/³⁄₄ cup) caster (superfine)
  sugar
185 g (6¹⁄₂ oz/1¹⁄₂ cups) self-raising flour,
  sifted
2 eggs, lightly beaten
3 tablespoons vegetable oil
3 tablespoons milk
100 g (3¹⁄₂ oz/heaped ³⁄₄ cup) grated
  dark chocolate
90 g (3¹⁄₄ oz/³⁄₄ cup) walnuts, finely
  chopped
whipped cream, to serve (optional)

Preheat the oven to 180°C (350°F/Gas 4). Lightly grease a 20 x 10 cm (8 x 4 inch) loaf (bar) tin and line the base with baking paper.

Put the mashed banana and sugar in a large bowl and mix well. Add the flour, eggs, oil and milk and gently stir until well combined. Stir in the chocolate and walnuts.

Pour the mixture into the prepared tin and bake for 55 minutes, or until a cake tester inserted into the centre of the cake comes out clean.

Remove from the oven and leave to cool in the tin for 5 minutes, then turn out onto a wire rack.

Serve warm, with whipped cream if desired.

NOTE: In warm weather, chocolate can be grated more easily if left to harden in the freezer for a few minutes before grating.

BANANAS FOSTER

SERVES 4

2 tablespoons unsalted butter
4 firm, ripe bananas, sliced in half
    lengthways
2 tablespoons soft brown sugar
2 tablespoons rum
vanilla ice cream, to serve

Melt the butter in a large frying pan. Add the banana halves, in batches if necessary, and briefly cook over medium–high heat, gently turning them to coat in butter.

Add the sugar and cook for 1 minute, or until the banana is caramelised.

Sprinkle with the rum, then divide among serving bowls and serve with a scoop of vanilla ice cream.

APPLE SAGO PUDDING

SERVES 4

4 tablespoons caster (superfine) sugar
100 g (3^1/$_2$ oz/^{1}/$_2$ cup) pearl sago
600 ml (21 fl oz) milk
55 g (2 oz/^{1}/$_2$ cup) sultanas (golden raisins)
^{1}/$_4$ teaspoon sea salt
1 teaspoon natural vanilla extract
a pinch of ground nutmeg
^{1}/$_4$ teaspoon ground cinnamon
2 eggs, lightly beaten
3 small braeburn or golden delicious
    apples (about 250 g/9 oz), peeled,
    cored and very thinly sliced
1 tablespoon soft brown sugar
whipped cream or ice cream, to serve

Preheat the oven to 180°C (350°F/Gas 4). Grease a 1.5 litre (52 fl oz/6 cup) ceramic soufflé dish.

Put the caster sugar, sago, milk, sultanas and sea salt in a saucepan. Bring to the boil, stirring often, then reduce the heat and simmer for 5 minutes.

Stir in the vanilla, nutmeg, cinnamon, eggs and apple slices, then gently pour into the soufflé dish.

Sprinkle with the brown sugar and bake for 45 minutes, or until the pudding is just set in the middle and golden brown on top.

Serve hot or warm, with whipped cream or ice cream.

Bananas foster

herbs and leaves

Herbs and salad leaves bring pep, crunch and zing to our tables. From the freshness of mint, the spicy depth of basil or the grassy familiarity of parsley, there's a herb for just about any dish. Herbs have fascinating histories, and most were first used medicinally before finding their way into the kitchen. Salad leaves are no less nutritious — or delicious! — bursting with vitality and goodness.

When choosing herbs or salad leaves, select young rather than older leaves, which have a coarser flavour and texture. Where possible, avoid cutting leaves as this crushes cells and causes unsightly bruising, which can affect flavour; when cutting is necessary, use a very sharp knife to minimise damage. Only wash or prepare herbs or leaves just before using or they will lose flavour and may even discolour. Any leaves that are a little wilted can be revived by being soaked briefly in a large bowl of iced water, then dried off well.

Basil
A member of the mint family, basil is most likely native to India, where it is considered sacred. Basil has a long medicinal history and is steeped in folkore: it was once believed inhaling its aroma would bring a scorpion into the brain!

Basil has a strong, spicy, anise flavour. Of all the basils, **sweet basil** is the sweetest-tasting and most common. It has rounded, tender leaves with a tapered end and flourishes in high summer. The pretty **dark opal** (or **red rubin**) basil has small, reddish-purple leaves that are slightly ruffled around the edge. It has assertive notes of clove and loses its vibrant colour when cooked, so is often used raw in salads. **Lemon, clove** and **anise basil** contain organic compounds which give them various flavour notes, hence their names. They can be used in much the same way as sweet basil, but are too overpowering for making a pesto. Many basils are used in Asian cooking, most notably **Thai** (or **holy**) **basil**, which has a pronounced spicy menthol flavour.

Avoid bunches with bruised or wilted leaves or tiny white blooms, as the flavour won't be so good. Store basil, still on its stems, in a tightly sealed plastic bag in the refrigerator for 2 days maximum.

Basil can be used raw, or added to hot dishes just before serving. It is fabulous with egg, pasta and rice dishes, tomato-based stews, sauces and soups, Mediterranean vegetable dishes, red wine, citrus, garlic and cheese.

BAY LEAVES

Bay leaves are the aromatic leaves of several types of laurel, the most common being *Laurus nobilis*. In ancient times, laurel wreaths were used to honour scholars, poets and Olympians. With their distinctive, slightly peppery flavour, bay leaves are a vital component of bouquet garni. Commercially dried leaves can taste musty and uninteresting, so use fresh leaves (or dried fresh leaves) when you can as their flavour is much stronger and sweeter.

If you have excess fresh bay leaves, remove them from their stems, tie in small bunches and hang upside down in a cool, dark place for 2–3 days until completely dry. Store in an airtight container and use within a few months, or freeze in small airtight plastic containers or bags for up to 6 months.

Bay leaves are added whole to dishes as they are not suited to chopping; they are generally removed before serving. Bruising the leaf will help release the fragrance. A single fresh leaf will infuse an entire potful of soup or stew.

Bay leaves enhance hearty flavours such as onion, bacon, most meats, red wine, tomatoes, mushrooms, root vegetables and pulses.

CHIVES

Related to the onion, the chive grows in clumps and has been cultivated for some 5000 years. Chives are very fragile and do not keep well once cut. They are sold in bunches and should be an even, deep green colour with no yellowing or wilting. Wrap damp paper towels around the root end and store in a zip-lock bag in the refrigerator; they should keep for 3 days.

Chives go beautifully in most egg dishes and salads, and also brighten potato dishes, including mash. They are good too with smoked salmon (and fish generally), in cream-based sauces for chicken or vegetables, and savoury baked items such as scones, muffins and cheesy breads.

CORIANDER (CILANTRO)

Coriander is a member of the carrot family and related to flat-leaf (Italian) parsley, which it vaguely resembles. This ancient herb was one of the 'bitter herbs' of the Passover meal. Both the dried seeds and fresh leaves and stems are used. In taste, the leaves are quite citrusy and aromatic, but as heat really kills the flavour, they are chopped and added to dishes at the last moment.

Look for bright-green coriander with unblemished leaves and the roots still attached. It doesn't store well, so keep it in an airtight container or plastic bag in the refrigerator for 1–2 days only.

Coriander especially complements the bright, fragrant, acidic flavours of Mexican, Asian and North African cuisines.

DILL

With its delicate, feathery, frond-like leaves and subtle aniseed flavour, dill is a lovely herb. Its name may derive from the Norse word *dilla*, which means to lull or soothe — and indeed dill has long been used to cure ailments from sleeplessness to colic. Often confused with the leafy ends of fennel, dill (also called dill weed) belongs to the parsley family. Dill leaves and seeds are an important ingredient in pickles (especially cucumber).

Fresh dill is sold in bunches — look for sprightly bright-green leaves. Wrap gently in damp paper towels inside a loosely sealed plastic bag and store it in the crisper. It should last 3–4 days; wilting won't affect the flavour.

Dill is divine with fish and seafood and pairs well with sour cream, yoghurt, mayonnaise, eggs, beetroot (beets), cabbage, chicken (particularly poached) and cucumber. Also try it in omelettes, salads and potato dishes.

LETTUCE

Part of the daisy family, lettuce is one of the world's most widely eaten vegetables and comes in many forms and colours. The leaves can be crisp and crinkly, or buttery, smooth and tender. Some lettuces have an alluring bitter edge while others are hot and peppery. Of the crisphead lettuces, **iceberg** is probably most common. This family is usually round, with layers of juicy, crisp, tightly packed leaves. Butterhead lettuces, of which **butter lettuce** is best known, have smaller heads with soft, more loosely packed leaves. Butterhead lettuces can be red or green; **mignonette**, **bibb** and **boston** are other examples. **Cos (romaine)** lettuce has an elongated shape with long, pointed leaves in a loosely packed head. The leaves are very crisp, with a sweet, nutty flavour. Loose-leafed lettuces include **lollo rosso** and **oakleaf**. A varied bunch, these particularly pretty lettuces have mild and delicate leaves that can be frilly, ruffled or smooth; some are green and others reddish.

Buy the freshest-looking lettuce and store it in a perforated plastic bag in the crisper for up to 2 days. Just before using, fill a sink with cool water, briefly dunk the leaves in the water to remove grit or bugs, then gently shake the leaves dry, or dry in a salad spinner. Dressings won't cling to wet leaves.

Lettuce is generally eaten raw, although the French have a famous braised lettuce dish, and the Greeks add lettuce to certain meat stews. The crispheads and cos (romaine) lettuce are best for these uses. In braises, soups or stews, lettuce goes well with dried beans, lamb, dill, thyme, mint, chicken stock, butter, olive oil, lemon and peas.

MARJORAM AND OREGANO

Part of the mint family, these perennials are very closely related; oregano is often called 'wild marjoram', while botanists now consider marjoram a type of oregano, of which there are about 50. They are associated with the Mediterranean, although types of oregano are also used in Latin America.

Marjoram has smaller, sweeter and softer leaves than oregano, which is quite spicy and pungent. These herbs will keep in the fridge for about a week, wrapped in damp paper towels in a loosely sealed plastic bag.

Oregano is excellent in any tomato-based dish, with fish, egg dishes and grilled meats (it is wonderful in a marinade), dried bean dishes, soups, stews and summer vegetables. It is legendary with lamb, lemon and garlic.

Marjoram is similarly superb in salad dressings and meat marinades, dried bean and egg dishes, and has a special affinity with poultry.

MINT

A lovely kitchen herb, mint's fresh, light flavour adds zing to so many dishes. It is thought the Romans introduced mint to England, where it famously became popular paired with lamb as mint sauce. In European cookery it is also widely used with potatoes or peas. It is used extensively in Middle Eastern, Asian and Indian cooking.

There are hundreds of varieties, the most common being **spearmint** and **peppermint**. Spearmint has long, slim, pointed leaves; peppermint has rounded, squatter leaves and a more intense mint taste.

Mint's soft, tender leaves are easily damaged and can quickly wilt. For the best flavour, select the very freshest mint you can find as its flavour quickly diminishes. The leaves should be sprightly, erect and bright green. Sealed in a plastic bag or container in the crisper it should keep for 3–4 days.

Mint is excellent in salads and with vegetables generally, as well as yoghurt, sour cream, vinegar, capers, olives, dried tomatoes, feta and ricotta cheese, lemon, orange, basil and even pineapple and strawberries.

PARSLEY

Parsley may well be the cooking world's most used herb. A Mediterranean native, its name means 'rock celery', and parsley is indeed related to celery. All herbs have nutrients but few are as rich in vitamins and minerals as parsley.

Two main types are used in the kitchen. **Curly parsley** has clusters of tight leaves with a strong flavour; the leaves become quite tough as it ages. It requires careful washing to remove dirt trapped in its leaves. **Flat-leaf (Italian) parsley** has flat leaves that become quite large. Unless specified as flat-leaf, the recipes in this book use curly parsley, but the two are interchangeable.

Older parsley develops a coarse taste. The curly type should have tight, compact leaves, with no yellowing; flat-leaf parsley should be bright green with neat, smallish leaves. Wrap the stalks in damp paper towels and store in a loosely sealed plastic bag in the crisper. Parsley should keep for 4–5 days.

Parsley is very high in chlorophyll, a green pigment that can bleed and ruin the appearance of pale dishes, such as those containing béchamel sauce, so it is best to chop and add parsley to such dishes just before serving.

Parsley complements all meats, fish and seafood, practically every vegetable you can think of, as well as cheeses, eggs and grains.

RADICCHIO

A type of chicory, radicchio's pleasantly earthy bitterness and gorgeous magenta colour add a lovely accent to green salads. Radicchio begins life green; its red colour is achieved by being grown in dark sheds. Its bitter flavour is due to a chemical called intybin, which stimulates the digestion.

Four commonly available varieties are the garnet-red, round **chioggia**, the elongated, looser-leafed and highly prized **treviso**, the soft, speckle-leafed **castelfranco**, and **radicchio di Verona**, which is slightly elongated and has bright maroon leaves. All take their names from towns in northern Italy.

Leaves should look crisp, and the stem end light-coloured and clean. Radicchio will keep in a perforated plastic bag in the crisper for 2–3 days.

Discard any tough outer leaves. Carefully wash and dry the remaining leaves just before use. To include them in a salad, tear (rather than cut) larger leaves into smaller pieces. For grilling (broiling) or braising, trim the outer leaves, then cut the whole head in half or quarters lengthways.

ROCKET (ARUGULA)

Rocket, also called rucola, has long been used in Italy as a salad green. Like its relatives radish and watercress, it has a strong, peppery flavour, and like most other salad greens is high in vitamins C and A, calcium and iron. Regular rocket has large leaves, which are used whole or roughly torn, with a thickish, tough lower stem that needs removing. 'Wild' rocket has tiny leaves with a very strong flavour; they require no trimming and are used whole.

Rocket is highly perishable — choose bright green leaves and store in a loosely sealed bag in the refrigerator for 2 days maximum.

A combination of good balsamic vinegar and olive oil is the best dressing for rocket-based salads. Rocket can also be used in hot dishes such as pasta, risottos, in tart and pie fillings, and raw sauces such as pesto.

ROSEMARY

Associated with remembrance, rosemary is a member of the mint family and native to southern Europe. It has a long culinary and medicinal history. Take care when cooking with rosemary as its pungency can easily overpower.

Select bunches with perky green leaves that smell very resinous when rubbed. Rosemary will stay fresh in an airtight bag or container in the refrigerator for up to 1 week.

Rosemary enhances digestion, and is brilliant in meat marinades. It is also great in soups and braises, particularly those involving Mediterranean vegetables, dried beans, tomatoes, lamb, lemon, potato and mushroom.

SAGE

There are over 500 types of this small evergreen shrub with pretty grey-green leaves. The ancient Greeks and Romans credited sage with all manner of healing properties; its Latin name, *salvia*, means 'to cure'.

Sage is rarely used as a salad herb as its flavour is too strong and woodsy, although deep-fried sage leaves are a delicious Tuscan antipasti. It is sold in small bunches and keeps refrigerated in an airtight container for up to 5 days.

Sage is excellent with rich meats and is believed to aid in their digestion. It is used in meat marinades, stuffings (especially for poultry) and in pork sausages and pâtés. It goes very well in deep, wintery dishes and with cheese (blue, parmesan, cheddar), onion, potato, lemon and butter.

THYME

A member of the mint family, this small, shrubby evergreen plant has a sweet, slightly spicy aroma and flavour. There are over 300 types, but the most common are **English thyme**, the herb most generally sold as 'thyme', and **lemon thyme**, which has a distinctive lemony smell and flavour.

Buy thyme with perky green leaves. Wrap the stems in damp paper towels and keep in a loosely sealed bag in the crisper for up to 1 week.

English thyme is lovely in dishes containing red meats, tomato and eggs. With its citrusy taste, lemon thyme is superb with fish and chicken.

WATERCRESS

Watercress is a fast-growing, semi-aquatic plant from the mustard family. An ancient Persian chronicler once noted greatly improved health in soldiers who were fed cress. Nowadays we know it is high in vitamins C, A and K, and also contains iron, calcium, folic acid and anti-cancer compounds.

Watercress has a peppery mustard flavour which becomes stronger as it matures and less pronounced when cooked. It is very delicate and perishable, so inspect bunches carefully. Avoid those with thick stems, which will taste very peppery. Sealed in a plastic bag, it will keep for 2 days in the crisper.

If using cress in salads, pick the sprigs and discard the stalks (the stalks can be added to cooked dishes along with the leaves). Only wash watercress just before using, then carefully dry in a salad spinner or pat dry with paper towels. Slightly wilted cress can be revived in a bowl of iced water.

Watercress goes well with meats (including cured meats) and fish (including smoked and pickled fish), with other salad greens, lemon, orange, walnuts, hazelnuts, avocado, onion, eggplant (aubergine), tomato, potato, sweet potato, olives, balsamic vinegar, and parmesan and blue cheese.

WITLOF (CHICORY/BELGIAN ENDIVE)

Related to radicchio, witlof is a member of the daisy family. Its production was only perfected in the 1800s in Belgium, hence its association with that country. The word 'witlof' is from *witloof*, meaning 'white leaf' in Flemish.

Witlof has a compact, slightly pointed shape and pale leaves, which result from lack of exposure to sunlight. Its growing conditions are labour intensive, so witlof is never cheap to buy. The flavour is sweet and almost creamy, with a subtle, bitter edge. There is also a very pretty red-tinged witlof, which tastes identical to the white one.

Look for very tightly closed, white heads. The greener the leaves, the more bitter they are, so the palest witlof is the most desirable. Make sure there are no blemishes or bruises. Witlof will keep for 2 days only, in a loosely sealed plastic bag in the crisper. Only wash and prepare witlof just before using as it goes grey once cut and does not keep well when wet.

To use raw in salads, cut off the bitter base, separate the leaves, give the quickest rinse in water and carefully shake dry. For cooking, trim the base and cut the witlof in half lengthways for braising, or into chunks for sautéeing. Never cook it in a cast-iron pan as it turns an awful grey colour.

In salads, witlof goes well with orange, walnuts, hazelnuts, blue cheese, apple, pear, creamy dressings, mild vinaigrettes, seafood and avocado. Cooked witlof is lovely with cream, butter, chives, parsley, cheese (blue and gruyère particularly) and citrus.

ROCKET TARTS
MAKES 4

2 sheets of frozen butter puff pastry,
  thawed
1 tablespoon olive oil
$1/2$ small onion, finely diced
a large handful of baby rocket (arugula)
  leaves
3 eggs, lightly beaten
125 g ($4^{1}/_{2}$ oz/$1/_{2}$ cup) ricotta cheese
a pinch of ground nutmeg

Preheat the oven to 180°C (350°F/Gas 4).

Cut four 15 cm (6 inch) circles from the puff pastry
and use them to line four greased 10 cm (4 inch) loose-based
tartlet tins. Prick the bases with a fork. Line the pastry shells
with baking paper and half-fill with baking beads, dried beans
or rice. Bake the pastry for 15 minutes, then remove the paper
and baking beads and bake for a further 5 minutes, or until
light golden. Remove from the oven and set aside.

Heat the olive oil in a frying pan. Add the onion and
sauté over medium heat for 5 minutes, or until softened. Add
the rocket and remove from the heat.

In a small bowl, combine the eggs, ricotta and nutmeg.
Season with sea salt and freshly ground black pepper and stir
to just combine; the ricotta will not be completely smooth.
Stir in the rocket mixture.

Spoon the filling into the pastry shells and bake for
25 minutes, or until set. Serve warm or at room temperature.

Rocket tarts are best eaten the day they are made.

HERB BAKED RICOTTA
SERVES 6–8

1 kg (2 lb 4 oz) firm, fresh ricotta
  (see Note)
2 tablespoons thyme leaves
2 tablespoons chopped rosemary
2 tablespoons chopped oregano
3 tablespoons chopped parsley
3 tablespoons snipped chives
2 garlic cloves, crushed
2 teaspoons freshly cracked black pepper
125 ml (4 fl oz/$1/_{2}$ cup) olive oil
crusty bread, to serve

Pat the ricotta with paper towels to absorb any excess liquid
and place in a baking dish.

In a bowl, mix together the herbs, garlic, pepper and
olive oil. Spoon the mixture onto the ricotta, gently pressing it
in with the back of the spoon. Cover and refrigerate overnight.

Preheat the oven to 180°C (350°F/Gas 4). Bake the
ricotta for 30 minutes, or until golden. Serve with crusty bread.

NOTE: If you can't buy a wedge of firm, fresh ricotta, drain the ricotta
in a colander overnight over a large bowl. Spread half the herb mixture
in a 1.25 litre (44 fl oz/5 cup) loaf (bar) tin, then spoon the ricotta in and
spread with the remaining herbs before baking.

SPAGHETTI WITH ROCKET AND CHILLI
SERVES 4–6

500 g (1 lb 2 oz) spaghetti or spaghettini
2 tablespoons olive oil
2 teaspoons finely chopped small red
    chilli
450 g (1 lb) rocket (arugula) leaves
1 tablespoon lemon juice
shaved parmesan cheese, to serve

Cook the pasta in a large saucepan of rapidly boiling salted water until *al dente*. Drain and return to the pan.

Meanwhile, heat the olive oil in a large frying pan. Add the chilli and cook for 1 minute over low heat. Add the rocket and cook, stirring often, for a further 2–3 minutes, or until softened. Add the lemon juice and season with sea salt and freshly ground black pepper.

Add the rocket mixture to the pasta and toss to combine well. Serve scattered with shaved parmesan.

LINGUINE PESTO
SERVES 4–6

2 large handfuls of basil leaves
2 garlic cloves, crushed
3 tablespoons pine nuts, toasted
185 ml (6 fl oz/3/4 cup) olive oil
50 g (1 3/4 oz/1/2 cup) grated parmesan
    cheese, plus extra shaved or grated
    parmesan, to serve
500 g (1 lb 2 oz) linguine

Put the basil, garlic and pine nuts in a food processor and blend until coarsely chopped. With the motor running, add the olive oil in a steady stream until mixed to a smooth paste. Transfer to a bowl, stir in the parmesan and season to taste with sea salt and freshly ground black pepper.

Cook the pasta in a large saucepan of rapidly boiling salted water until *al dente*. Drain and return to the pan.

Toss enough of the pesto through the pasta to coat well. Serve scattered with parmesan.

NOTE: Spoon any left-over pesto into an airtight jar, cover with a layer of olive oil and refrigerate for up to 1 week. The pesto can also be frozen for up to 1 month.

Spaghetti with rocket and chilli

ROAST MONKFISH WITH ROSEMARY AND GARLIC

SERVES 4

4 x 250 g (9 oz) skinless monkfish
   tail fillets, or any firm, white-fleshed
   fish fillets
3 large garlic cloves, sliced into
   thin slivers
24 small rosemary sprigs
6 slices of bacon, cut in half
4 tablespoons olive oil
lemon wedges, to serve

Preheat the oven to 200°C (400°F/Gas 6).
   Using a small sharp knife, make six small incisions in each fish fillet and insert a sliver of garlic and a small sprig of rosemary into each one. Season the fish with sea salt and freshly ground black pepper and wrap some bacon around each fillet.
   Place the fish in a roasting tin and drizzle the olive oil over the top. Roast for 15 minutes, or until the fish is cooked through. Serve with lemon wedges.

SALMORIGLIO

MAKES ABOUT 150 ML (5 FL OZ)

2 tablespoons marjoram
125 ml (4 fl oz/½ cup) extra virgin
   olive oil
1 tablespoon lemon juice

Pound the marjoram using a mortar and pestle. Transfer to a bowl and gradually add the olive oil and lemon juice. Season to taste with sea salt and freshly ground black pepper.
   Salmoriglio can be kept covered in the refrigerator for several days. Bring to room temperature before serving.

NOTE: This simple dressing is widely used in Italy to accompany grilled fish or other seafood. Thyme or oregano can be used instead of marjoram.

PORK WITH SAGE AND CAPERS

SERVES 4

25 g (1 oz) unsalted butter

3 tablespoons extra virgin olive oil

1 onion, finely chopped

100 g (3½ oz/1¼ cups) fresh white
  breadcrumbs

2 teaspoons chopped sage, plus 8 whole
  sage leaves, to garnish

1 tablespoon chopped flat-leaf (Italian)
  parsley

2 teaspoons grated lemon zest

2½ tablespoons salted baby capers,
  rinsed and drained

1 egg, lightly beaten

2 large pork fillets, about 500 g (1 lb 2 oz)
  each

8 large, thin slices of bacon or prosciutto

2 teaspoons plain (all-purpose) flour

100 ml (3½ fl oz) dry vermouth

310 ml (10¾ fl oz/1¼ cups) chicken or
  vegetable stock

Preheat the oven to 170°C (325°F/Gas 3).

Heat the butter and 1 tablespoon of the olive oil in a frying pan. Add the onion and sauté over medium heat for 5 minutes, or until lightly golden. Place in a bowl with the breadcrumbs, chopped sage, parsley, lemon zest, ½ tablespoon of the capers and the egg. Season well with sea salt and freshly ground black pepper and mix together well.

Slice each pork fillet in half lengthways, taking care not to cut all the way through, then open the pork out. Spread the breadcrumb mixture over one piece of pork fillet, then cover with the other fillet.

Lay the bacon slices lengthways on a work surface, overlapping the slices so there are no gaps. Place the pork lengthways down one side of the bacon slices, then carefully roll the bacon and pork up, taking care the pork is completely covered. Tie the roll with kitchen string at 2.5 cm (1 inch) intervals to secure the bacon.

Place the pork in a flameproof baking dish and drizzle with 1 tablespoon of the olive oil. Bake for 1 hour. To test if the meat is cooked, insert a skewer in the thickest part — the juices should run clear. When the pork is cooked, remove from the baking dish, cover with foil and leave to rest.

Place the baking dish on the stovetop over medium heat. Add the flour and stir in well. Stir in the vermouth and allow to bubble for 1 minute. Pour in the stock and stir well to remove any lumps, then simmer for 5 minutes. Stir in the remaining capers.

In a small saucepan, heat the remaining oil. When the oil is very hot, fry the whole sage leaves until crisp. Remove and drain on paper towels.

Slice the pork into 1 cm (½ inch) strips. Serve drizzled with the sauce and garnished with the fried sage leaves.

GRILLED SARDINES WITH BASIL AND LEMON
SERVES 4

1 lemon
12 whole sardines, cleaned and scaled
coarse sea salt, for seasoning
4 tablespoons olive oil
3 tablespoons torn basil leaves or
   small whole leaves

Preheat the oven grill (broiler) to very hot.

Thinly slice the lemon, then cut each slice in half. Insert several pieces inside each sardine. Season both sides of each sardine with the sea salt and some freshly ground black pepper.

Put the sardines on a baking tray, drizzle with half the olive oil and grill for 3 minutes on each side, or until just cooked through; the flesh inside the sardines should be opaque. Remove and place in a shallow serving dish.

Scatter the basil over the sardines and drizzle with the remaining olive oil. Serve warm or at room temperature.

VARIATION: Small herring or mackerel can be used instead of sardines.

CORIANDER PORK WITH FRESH PINEAPPLE
SERVES 4

400 g (14 oz) pork loin or fillet, trimmed
¼ pineapple
1 tablespoon vegetable oil
4 garlic cloves, chopped
4 spring onions (scallions), chopped
1 tablespoon fish sauce
1 tablespoon lime juice
a large handful of coriander (cilantro) leaves
a large handful of chopped mint
steamed rice, to serve

Partially freeze the pork until it is just firm, then slice thinly. Cut the skin off the pineapple, then cut the flesh into bite-sized pieces.

Heat the oil in a wok or heavy-based frying pan. Add the garlic and spring onion and cook over medium–high heat for 1 minute. Remove from the wok.

Heat the wok to very hot, then add the pork in batches and stir-fry for 2–3 minutes, or until just cooked.

Return the garlic, spring onion and all the pork to the wok and add the pineapple, fish sauce and lime juice. Toss together, then cook for 1 minute, or until the pineapple is heated through.

Toss the coriander and mint through and serve immediately, with steamed rice.

MUSHROOM QUICHE WITH PARSLEY PASTRY

SERVES 4–6

155 g (5½ oz/1¼ cups) plain
   (all-purpose) flour
3 tablespoons very finely chopped parsley
90 g (3¼ oz) cold unsalted butter,
   chopped
1 egg yolk, mixed with 2 tablespoons
   iced water

MUSHROOM FILLING
30 g (1 oz) unsalted butter
1 red onion, finely chopped
175 g (6 oz) button mushrooms, sliced
1 teaspoon lemon juice
4 tablespoons chopped parsley
3 tablespoons snipped chives
2 eggs, lightly beaten
170 ml (5½ fl oz/⅔ cup) pouring
   (whipping) cream

Sift the flour and a pinch of sea salt into a large bowl. Mix the parsley through. Using your fingertips, lightly rub the butter into the flour until the mixture resembles coarse breadcrumbs. Make a well in the centre. Add the egg yolk mixture to the well and mix using a flat-bladed knife until a rough dough forms, adding a little extra iced water if needed. Turn out onto a lightly floured surface and gather into a ball. Cover with plastic wrap and refrigerate for at least 30 minutes.

Roll out the pastry on a sheet of baking paper until large enough to fit the base and side of a 35 x 10 cm (14 x 4 inch) loose-based tart tin. Ease the pastry into the tin and trim the edges. Refrigerate the pastry-lined tin for a further 20 minutes.

Meanwhile, preheat the oven to 190°C (375°F/Gas 5).

Line the pastry shell with baking paper and spread with a layer of baking beads, dried beans or rice. Bake the pastry for 15 minutes, then remove the paper and baking beads and bake for a further 10 minutes, or until the pastry is dry. Reduce the oven temperature to 180°C (350°F/Gas 4).

To make the mushroom filling, melt the butter in a frying pan, add the onion and sauté over medium heat for 5 minutes, or until softened. Add the mushrooms and sauté for 2–3 minutes, or until soft. Stir in the lemon juice and herbs. Meanwhile, mix the eggs and cream together and season with sea salt and freshly ground black pepper.

Spread the mushroom mixture into the pastry shell and pour the egg mixture over. Bake for 25–30 minutes, or until the filling has set. Serve warm or at room temperature.

Mushroom quiche is best served the day it is made.

INVOLTINI OF SWORDFISH

SERVES 4

1 kg (2 lb 4 oz) swordfish, sliced as thinly as possible (see Note)

3 lemons, plus extra lemon wedges, to serve

4 tablespoons olive oil

1 small onion, chopped

3 garlic cloves, chopped

2 tablespoons capers, rinsed, drained and chopped

2 tablespoons finely chopped pitted kalamata olives

4 tablespoons grated parmesan cheese

120 g (4¼ oz/1½ cups) soft white breadcrumbs

2 tablespoons chopped flat-leaf (Italian) parsley

1 egg, lightly beaten

24 fresh bay leaves

2 small white onions, quartered and separated

Soak eight wooden skewers in cold water for 20 minutes to prevent scorching.

Remove the skin from the swordfish slices if necessary and cut each slice in half. Cut away the very dark flesh. Place each piece between two sheets of plastic wrap and roll gently with a rolling pin to flatten the fish without tearing it. Cut each piece into two 4 x 10 cm (1½ x 4 inch) strips.

Using a vegetable peeler, thinly peel the rind from the lemons, then cut the rind into 32 even pieces. Squeeze the lemons to give 3 tablespoons juice. Whisk the lemon juice with 2 tablespoons of the olive oil and set aside for basting the fish.

Heat the remaining olive oil in a frying pan. Add the chopped onion and garlic and sauté over medium heat for 2–3 minutes, or until softened slightly. Transfer to a bowl, add the capers, olives, parmesan, breadcrumbs and parsley and mix together. Season to taste with sea salt and freshly ground black pepper, then add the egg and mix to combine well.

Divide the stuffing among the fish pieces. Using lightly oiled hands, roll up the fish to form neat rolls. Thread four rolls onto each skewer, alternating with the bay leaves, lemon rind pieces and the small white onion quarters.

Heat a barbecue grill plate or chargrill pan to medium–high. Give the reserved basting mixture a stir. Barbecue or grill the swordfish skewers for 3–4 minutes on each side, basting regularly.

Serve hot, with lemon wedges.

NOTE: Ask your fishmonger to slice the swordfish very thinly for you — the slices should only be about 7–8 mm (3/8 inch) thick.

GREEN SALAD WITH LEMON VINAIGRETTE

SERVES 6

1 baby cos (romaine) lettuce
1 small butter lettuce
50 g (1³⁄₄ oz/1²⁄₃ cups) watercress sprigs
100 g (3¹⁄₂ oz) rocket (arugula) leaves

LEMON VINAIGRETTE
1 tablespoon finely chopped French
  shallot
2 teaspoons dijon mustard
¹⁄₂ teaspoon sugar
1 tablespoon finely chopped basil
1 teaspoon grated lemon zest
3 teaspoons lemon juice
1 tablespoon white wine vinegar
1 teaspoon lemon oil (see Note)
4 tablespoons extra virgin olive oil

Discard the outer leaves from the lettuces, then separate the inner leaves. Wash the lettuce, watercress and rocket, then drain and dry thoroughly. Place in the refrigerator to chill.

To make the lemon vinaigrette, put the shallot, mustard, sugar, basil, lemon zest, lemon juice and vinegar in a bowl and whisk together well. Mix the lemon and olive oils in a small pouring jug and slowly add to the bowl in a thin stream, whisking constantly until a smooth, creamy dressing forms. Season to taste with sea salt and freshly ground black pepper.

Put the salad greens in a large bowl. Just before serving, drizzle the vinaigrette over the salad and toss gently to coat.

NOTE: Lemon oil is available from gourmet food stores, or you can make your own by steeping some lemon zest in some extra virgin olive oil. The flavour will become more intense the longer it steeps. Alternatively, just add 1 teaspoon finely grated lemon zest to the vinaigrette and omit the lemon oil.

DILL COLESLAW

SERVES 6–8

4 tablespoons sour cream
2 teaspoons grated fresh horseradish
  (see Note)
1 tablespoon lemon juice
1 tablespoon dijon mustard
2 tablespoons finely chopped dill
300 g (10¹⁄₂ oz/4 cups) finely shredded
  red cabbage
2 carrots, grated

Put the sour cream, horseradish, lemon juice, mustard and dill in a large bowl and stir to combine well. Add the cabbage and carrot, toss well to coat the vegetables with the dressing, then season to taste with sea salt and freshly ground black pepper. Cover and refrigerate until ready to serve.

NOTE: If fresh horseradish is unavailable, substitute bottled horseradish (or horseradish cream; see recipe on page 101) to taste.

Green salad with lemon vinaigrette

CAESAR SALAD
SERVES 4–6

DRESSING
3 eggs
3 garlic cloves, crushed
2–3 anchovy fillets, drained
1 teaspoon worcestershire sauce
2 tablespoons lemon juice
1 teaspoon dijon mustard
185 ml (6 fl oz/¾ cup) olive oil

3 slices of white bread
20 g (¾ oz) unsalted butter
1 tablespoon olive oil
3 slices of bacon
1 large or 4 baby cos (romaine) lettuces
75 g (2½ oz/¾ cup) shaved parmesan
cheese

To make the dressing, put the eggs, garlic, anchovies, worcestershire sauce, lemon juice and mustard in a food processor and blend until smooth. With the motor running, add the olive oil in a thin, steady stream and process until the mixture is well combined and creamy. Season to taste with sea salt and freshly ground black pepper. Cover the surface directly with plastic wrap to prevent a skin forming and set aside.

Remove the crusts from the bread and cut the bread into 1.5 cm (⅝ inch) cubes. Heat the butter and olive oil in a frying pan. Add the bread and cook over medium heat for 5–8 minutes, or until golden all over, stirring often. Remove the croutons from the pan using a slotted spoon and drain well on paper towels.

Cook the bacon in the same frying pan for 3 minutes, or until golden and crisp, turning once. Break the bacon into bite-sized pieces.

Discard the tough outer leaves from the lettuce and separate the inner leaves. Wash and drain well, then dry thoroughly. Place in a bowl, add the dressing and toss.

Mix in the croutons and bacon, scatter the parmesan over the salad and serve.

Sauteed witlof with olives, anchovies and caperberries

SERVES 4 AS A STARTER OR SIDE DISH

40 g (1 1/2 oz/ 1/4 cup) pitted kalamata
   olives, chopped
2 anchovy fillets, drained and chopped
6 small caperberries
1 tablespoon olive oil
2 pale cream or red witlof (chicory/
   Belgian endive)
20 g (3/4 oz) unsalted butter
1 garlic clove, crushed
a pinch of chilli flakes (optional)

Put the olives and anchovies in a small bowl. Chop two of the caperberries and add to the olives with half the olive oil. Stir to combine well.

Discard the outer leaves from each witlof and cut the heads in half lengthways. Open out the leaves a little and spoon the olive mixture between the leaves. Join the two halves together again and tie with kitchen string to secure.

Heat the remaining oil and the butter in a saucepan over low heat. Add the witlof, garlic and chilli flakes, if using, then cover and cook for 8–10 minutes, turning the witlof once and adding a little hot water if necessary to stop it sticking.

To serve, untie the string and divide the witlof, cut side up, among serving plates. Spoon over any pan juices. Slice the remaining caperberries in half lengthways and scatter them over the witlof. Serve hot.

CHARGRILLED RADICCHIO
SERVES 4 AS A SIDE DISH

2 radicchio
3 tablespoons olive oil
1 teaspoon balsamic vinegar

Trim the radicchio, discarding the outer leaves. Slice the heads into quarters lengthways and rinse well. Drain well, then pat dry with paper towels.

Heat a chargrill pan or barbecue chargrill plate to high. Lightly drizzle the radicchio with some of the olive oil and season with sea salt and freshly ground black pepper. Cook for 2–3 minutes, or until the outer leaves soften and darken, then turn to cook the other side. Transfer to a dish and sprinkle with the remaining oil and vinegar.

Serve hot with grilled meats, or at room temperature as part of an antipasti platter.

SORREL AND ASPARAGUS RISOTTO

SERVES 4

1 litre (35 fl oz/4 cups) vegetable or
  chicken stock
250 ml (9 fl oz/1 cup) dry white wine
50 g (1¾ oz) unsalted butter, chopped
3 tablespoons extra virgin olive oil
1 onion, finely chopped
2 garlic cloves, crushed
300 g (10½ oz/1⅓ cups) risotto rice
24 asparagus spears, trimmed and cut into
  2 cm (¾ inch) lengths
100 g (3½ oz) sorrel, washed, trimmed
  and finely chopped (see Note)
60 g (2¼ oz/⅔ cup) grated parmesan
  cheese

Pour the stock and wine into a large saucepan and bring to the boil. Reduce the heat, then cover and keep at simmering point.

Melt half the butter in a large heavy-based saucepan. Add the olive oil, onion and garlic and sauté over medium heat for 5 minutes, or until the onion has softened. Add the rice and stir to coat, then add 125 ml (4 fl oz/½ cup) of the simmering stock and cook, stirring constantly, over low heat until all the stock has been absorbed. Continue adding the stock, 125 ml (4 fl oz/½ cup) at a time, stirring constantly and making sure the stock has been absorbed before adding more.

When nearly all of the stock has been added, stir in the asparagus and sorrel. Continue adding the stock until the rice is tender and creamy and the asparagus is just cooked.

Stir in the remaining butter and the parmesan. Season with sea salt and freshly ground black pepper and serve.

NOTE: Sorrel is a herb with long, tender green leaves and a tart, lemony flavour that is much appreciated in Europe. Look for it in good greengrocer stores when it appears in spring. Discard the bitter stems before cooking with this herb. Sorrel has a strong flavour, so is best used judiciously.

CHICKEN AND WATERCRESS STRUDEL
SERVES 6

60 g (2 1/4 oz/3/4 cup) fresh white
  breadcrumbs
1–2 teaspoons sesame seeds
180 g (6 oz/6 cups) watercress sprigs
4 boneless, skinless chicken breasts
125 g (4 1/2 oz) unsalted butter
3 tablespoons dijon mustard
250 ml (9 fl oz/1 cup) thick
  (double/heavy) cream
15 sheets of filo pastry

Preheat the oven to 190°C (375°F/Gas 5). Spread the breadcrumbs and sesame seeds on separate baking trays and bake for 8–10 minutes, or until golden.

Steam the watercress for 3 minutes, or until just wilted, then drain well. When cool enough to handle, use your hands to squeeze out the excess water.

Slice the chicken into thin strips. Heat 25 g (1 oz) of the butter in a large frying pan over medium–high heat. Add the chicken and cook, stirring constantly, for 4–5 minutes, or until just cooked through.

Remove the chicken using a slotted spoon, then add the mustard and cream to the pan. Stir well, reduce the heat and simmer until reduced to 125 ml (4 fl oz/1/2 cup). Remove from the heat and stir in the chicken and watercress.

Put the remaining butter in a small saucepan and melt over low heat. Lay a sheet of filo pastry on a work surface and cover the remaining sheets with a damp tea towel (dish towel) so they don't dry out. Brush the pastry with some of the melted butter and sprinkle with the toasted breadcrumbs. Lay another sheet of pastry on top, brush with more butter and sprinkle with more breadcrumbs. Repeat with the remaining filo and breadcrumbs.

Place the chicken mixture along the centre of the pastry. Fold the sides over, then roll into a parcel. Place on a greased baking tray, seam side down. Brush with the remaining butter, sprinkle with the sesame seeds and bake for 30 minutes, or until the pastry is golden.

Remove from the oven and allow the strudel to cool slightly before serving.

WATERMELON, FETA AND WATERCRESS SALAD

SERVES 4

2 tablespoons sunflower seeds
1 kg (2 lb 4 oz) seedless watermelon
180 g (6 oz/1¼ cups) crumbled feta
  cheese
75 g (2½ oz/2½ cups) watercress sprigs
2 tablespoons olive oil
1 tablespoon lemon juice
2 teaspoons chopped oregano

Heat a small frying pan over medium–high heat. Add the sunflower seeds and, shaking the pan continuously, dry-fry for 2 minutes, or until lightly golden. Tip the seeds into a bowl so they don't burn.

Cut the rind away from the watermelon, then cut the flesh into thick wedges. Place in a large serving dish with the feta and watercress and toss gently to combine.

In a small bowl, whisk together the olive oil, lemon juice and oregano. Season to taste with a little sea salt and freshly ground black pepper (don't add too much salt as the feta is already quite salty).

Pour the dressing over the salad and gently toss together. Scatter with the toasted sunflower seeds and serve.

POTATO GNOCCHI WITH SAGE AND PANCETTA SAUCE
SERVES 4

1 kg (2 lb 4 oz) floury potatoes, such as russet (idaho) or king edward, unpeeled
200 g (7 oz/scant 1²/₃ cups) plain (all-purpose) flour, plus extra, for kneading

SAGE AND PANCETTA SAUCE
20 g (³/₄ oz) unsalted butter
80 g (2³/₄ oz) pancetta or bacon slices, cut into thin strips
8 very small sage or basil leaves
150 ml (5 fl oz) thick (double/heavy) cream
50 g (1³/₄ oz/¹/₂ cup) shaved parmesan cheese

Preheat the oven to 180°C (350°F/Gas 4).

Prick the potatoes all over, then place in a non-stick roasting tin and bake for 1 hour, or until tender. Leave to cool for 15 minutes, then peel and put through a potato ricer or food mill (do not use a blender or food processor or the potato will become gluggy).

Gradually stir the flour into the potato. When the mixture gets too firm to use a spoon, work the flour in with your hands. Once a loose dough forms, transfer to a lightly floured surface and knead gently. Work in just enough extra flour to give a soft, pliable dough that is damp to the touch but not sticky, taking care not to add too much flour or the gnocchi will be heavy.

Divide the dough into six even portions. Working with one portion at a time, roll each one out on a floured surface to make a rope about 1.5 cm (⁵/₈ inch) thick. Cut the rope into 1.5 cm (⁵/₈ inch) lengths. Take one piece of dough and press your finger into it to form a concave shape, then roll the outer surface over the tines of a fork to make deep ridges. Fold the outer lips in towards each other to make a hollow in the middle. Set aside and continue with the remaining dough.

Bring a large saucepan of salted water to the boil. Add the gnocchi in batches, making sure not to crowd the pot. Stir gently and cook for 1–2 minutes, or until they rise to the surface. Remove with a slotted spoon, drain and place in a greased shallow casserole dish or baking tray.

Meanwhile, preheat the oven to 200°C (400°F/Gas 6).

To make the sage and pancetta sauce, melt the butter in a frying pan and fry the pancetta until crisp. Remove and set aside on paper towels to drain. Briefly fry the sage leaves until crisp, then remove and drain on paper towels. Add the cream to the pan, season with sea salt and freshly ground black pepper and gently simmer for 5–10 minutes, or until thickened.

Pour the sauce over the gnocchi and gently mix through. Scatter with the parmesan, pancetta and sage and bake for 10–15 minutes, or until the cheese has melted and is golden. Serve immediately.

CHEESE AND CHIVE SCONES
MAKES 9

250 g (9 oz/2 cups) self-raising flour
30 g (1 oz) cold unsalted butter, chopped
60 g (2¼ oz/½ cup) grated cheddar
　　cheese, plus 3 tablespoons extra,
　　for sprinkling
3 tablespoons grated parmesan cheese
2 tablespoons snipped chives
125 ml (4 fl oz/½ cup) milk
butter, to serve

Preheat the oven to 210°C (415°F/Gas 6–7) and grease a baking tray.

Sift the flour and a pinch of sea salt into a bowl. Using your fingertips, lightly rub in the butter until the mixture resembles breadcrumbs. Stir in the cheddar, parmesan and chives and make a well in the centre. Add the milk and 125 ml (4 fl oz/½ cup) water to the well, then mix lightly with a flat-bladed knife until a soft dough forms, adding a little more milk if the mixture is too dry.

Turn the dough out onto a lightly floured surface and knead briefly, taking care not to overwork the dough or the scones will be tough. Gently press the dough out to a 2 cm (¾ inch) thickness. Use a floured 5 cm (2 inch) plain round cutter to cut nine rounds from the dough. Place close together on the baking tray and sprinkle with the extra cheddar.

Bake for 12 minutes, or until the cheese is golden and the scones have risen. Serve warm, with butter.

Cheese and chive scones are best eaten the day they are made.

MINT JELLY

FILLS THREE 250 ML (9 FL OZ/1 CUP) JARS

1 kg (2 lb 4 oz) green apples
125 ml (4 fl oz/$\frac{1}{2}$ cup) lemon juice
3 very large handfuls of mint leaves
750 g (1 lb 10 oz/3$\frac{1}{2}$ cups) sugar,
 approximately
2–3 drops of green food colouring

Wash the apples and cut into thick slices, without peeling or coring them. Place in a large heavy-based saucepan with the lemon juice, most of the mint and 1 litre (35 fl oz/4 cups) water. Bring to the boil, then reduce the heat and gently cook for 10 minutes, or until the apple is soft and pulpy. Break up any large pieces with a wooden spoon.

Strain the mixture through a piece of muslin (cheesecloth) into a bowl — do not press the liquid through or it will become cloudy. Leave at room temperature overnight.

Put two plates in the freezer. Measure the strained juice and pour it into a large heavy-based saucepan. Add 220 g (7$\frac{3}{4}$ oz/1 cup) sugar for each 250 ml (9 fl oz/1 cup) liquid. Stir over low heat without boiling until the sugar has dissolved completely.

Bring to the boil, then reduce the heat and simmer over low heat for 20 minutes, then start testing for setting point by placing a little of the hot jelly on a chilled plate. When setting point is reached, a skin will form on the surface and leave a trail when you push your finger through it. If the jelly doesn't set, keep cooking and testing until it does.

Finely chop the remaining mint and add to the jelly with the food colouring. Stir well. Remove from the heat and leave for 5 minutes, then pour into hot, sterilised jars (see Note below) and seal. Allow to cool, then label and date each jar.

Store in a cool, dark place for up to 6 months. Once opened, mint jelly will keep in the refrigerator for up to 6 weeks.

Mint jelly is delicious with roast lamb.

NOTE: Jars must always be sterilised before pickles, preserves or jams are put in them for storage, otherwise bacteria will multiply. To sterilise your jars and lids, rinse them with boiling water and place in a warm oven for 20 minutes, or until completely dry. (Jars with rubber seals are safe to warm in the oven and won't melt.) Never dry your jars with a tea towel (dish towel) — even a clean one may have germs on it and contaminate the jars.

Onion and Thyme Marmalade
Fills three 250 ml (9 fl oz/1 cup) jars

2 kg (4 lb 8 oz) onions, halved and
thinly sliced
750 ml (26 fl oz/3 cups) red or white
wine vinegar
6 black peppercorns
2 bay leaves
800 g (1 lb 12 oz/4⅓ cups) soft brown
sugar
2 tablespoons thyme leaves
1 teaspoon sea salt
10 thyme sprigs, each about 3 cm
(1¼ inches) long

Put the onion and vinegar in a large heavy-based saucepan. Place the peppercorns and bay leaves in a 10 cm (4 inch) square of muslin (cheesecloth) and tie securely with string. Add it to the saucepan and bring to the boil, then reduce the heat and simmer for 40–45 minutes, or until the onion is very soft.

Add the sugar, thyme leaves and sea salt and stir for 7–8 minutes, or until the sugar has dissolved. Bring to the boil, then reduce the heat and simmer for 20–30 minutes, or until the mixture is thick and syrupy. Using a slotted spoon, skim off any impurities that rise to the surface. Discard the muslin bag and stir in the thyme sprigs.

Pour into hot sterilised jars (see Note below) and seal. Allow to cool, then label and date each jar. Store in a cool, dark place for 6–12 months. Once opened, onion and thyme marmalade will keep in the refrigerator for up to 6 weeks.

Onion and thyme marmalade is delicious with beef, venison, sausages, sharp cheeses and as a sandwich relish.

NOTE: Jars must always be sterilised before pickles, preserves or jams are put in them for storage, otherwise bacteria will multiply. To sterilise your jars and lids, rinse them with boiling water and place in a warm oven for 20 minutes, or until completely dry. (Jars with rubber seals are safe to warm in the oven and won't melt.) Never dry your jars with a tea towel (dish towel) — even a clean one may have germs on it and contaminate the jars.

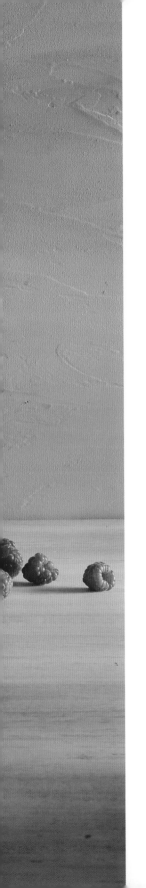

seasonal treats

Some produce, despite our best efforts, just can't be coaxed to grow out of season, or to travel very well. Those who have the patience and wisdom to wait are able to savour and celebrate the best that each month brings. While the natural season for the produce in this chapter is all too brief, any excess bounty can often be frozen or preserved to relish throughout the year.

ARTICHOKES

The artichoke is an unopened flower bud, and inside its formidable-looking armoury lies a tender, hidden heart. If left to bloom, it opens out into a spectacular large purple flower. Much of the artichoke itself is inedible, and there are those who believe they are far more trouble to prepare for the return they give. Interestingly, artichokes contain a chemical called cynarin, which reacts with other foods and wine, enhancing their flavour.

Although artichokes crop in spring and again in late autumn, they are not very high yielding, hence their expense. They vary greatly in size, from less than 100 g (3½ oz) up to 500 g (1 lb 2 oz) each.

Choose artichokes that feel heavy and have tight leaves (if they are starting to open they are past their prime). Generally the leaves should be green, but sometimes you might find artichokes with a slight bronzing; this is called 'winter bronzing' and is the effect of frost. It actually improves their flavour and many aficionados seek these out. Truly fresh artichokes make a characteristic 'squeaking' noise when their leaves are rubbed together.

Store artichokes in a sealed plastic bag or container in the refrigerator for 2–3 days maximum — and don't cut them, under any circumstances, until you are ready to use them. Cut artichokes very quickly turn brown on contact with air, so as you cut them, rub the cut surfaces with lemon juice, or slip them into a bowl of acidulated water to prevent browning.

To prepare artichokes, cut the stem flush with the base so they sit flat, or simply trim off the very end of the stem and peel the rest. Pull off the very tough outer leaves. Using kitchen scissors, cut off the top third of each remaining leaf. Steam, braise or gently boil the artichoke until tender (test by piercing the thickest part with a metal skewer). Depending on their size, they may take 10–30 minutes to cook. They can be served as they are with melted butter, hollandaise sauce, olive oil or a good home-made vinaigrette for dipping each leaf into, before eating the tender, lower part — finger bowls and serviettes are a necessary part of the ritual!

Artichokes are also served stuffed and baked, fried, marinated, preserved in oil or in a vegetable or delicate meat or fish stew. Their mellow, haunting flavour goes well with asparagus, peas, green beans, broad (fava) beans, chicken, veal, lamb, pine nuts, almonds, capers, prosciutto, lemon, olives, parsley, marjoram, thyme, and parmesan, gruyère and pecorino cheese.

Asparagus

After a dark, cold winter, you know there's sunshine around the corner when spears of asparagus first appear. Asparagus loses moisture and flavour from the moment it is picked and doesn't benefit from long-haul travel, so really, it is only worth eating locally grown asparagus during its brief season in spring.

The tip of an asparagus spear is an unopened bud which, if allowed to mature, grows into a fern-like plant with red berries. Green, purple and creamy-white are the most common types. Spears can be as thick as your thumb or as slender as a knitting needle, and debates rage as to which sort gives the superior eating experience. Actually, one asparagus plant can produce both fat or thin spears! Neither is superior to the other; which type you choose should be determined by how you will cook and serve them.

Reject spears with wrinkled or dried-out stem ends. The tips should be compact and bright, with no signs of damage or dampness. Choose firm spears of similar size so they'll take the same time to cook. Asparagus likes a little humidity, so put the spears upright in a container of water, cover well with plastic wrap and keep in the refrigerator, or store in a sealed plastic bag.

Before cooking, trim the woody ends by snapping them off where they naturally break, or trimming them with a knife. Thicker spears may have fibrous skin that needs to be peeled using a vegetable peeler; you may only need to peel the lower portion. Trimmings and peelings can be used to intensify the final asparagus flavour of soups and stocks destined for an asparagus dish (such as the stock for making an asparagus risotto).

Besides steaming or boiling to tenderness, asparagus can be doused in olive oil and roasted in a moderate oven for about 15 minutes, or barbecued over medium heat; these dry-cook methods concentrate flavours, making asparagus cooked this way ideal for robustly flavoured salads or antipasti dishes. Asparagus also makes great party food, either rolled up in a neat piece of bread, or wrapped in smoked salmon or prosciutto and served with a mild aïoli.

Asparagus is gorgeous on its own or with butter, hollandaise sauce, toasted breadcrumbs, garlic, seafood, smoked salmon, prosciutto, tarragon, dill, parsley, thyme, chives, anchovies, capers, olives, sun-dried tomatoes, zucchini (courgette), broad (fava) beans and artichoke.

Berries

Perhaps nothing expresses the season's bounty better than a basket brimming with berries. All berries are highly perishable, so feast on them quickly while they're at their freshest. They are also very fragile and prone to mould, so inspect carefully before buying. Only wash your berries just before you eat them — water impairs their flavour, and once damp, they speedily deteriorate.

Blackberries

Blackberries belong to the rose family and taste best when picked wild from their prickly bush — although those from the thornless cultivated ones are still delicious! Their season is late summer to autumn and they'll taste slightly sour unless picked fully ripe. Look for plump, shiny fruit, with deep black-purple skins and no stem; stems indicate they weren't quite mature when picked. If possible, use on the day of purchase, or freeze for up to 6 months.

Blackberries are lovely in compotes, baked desserts such as blackberry pie, and chilled desserts like jellies, creams and sorbets. They make incredible jam, pair well with game meats such as venison and pigeon, and are lovely with apples, other berries, cream, mascarpone cheese, crème fraîche, yoghurt, orange, cinnamon, red wine, sparkling wine and pastry.

Blueberries

With their smooth, indigo-blue skin and squat, round shape, these small, sweet-tasting North American natives are an attractive addition to fruit salads, cake fillings, pie toppings, cheesecakes and chilled desserts.

Blueberries should be lightly coated with a powdery, silver-white bloom, which signifies freshness and is their natural sun shield. Look for firm, plump, uniform-sized fruits with no signs of wrinkling. Store them for no longer than 2 days, preferably in a paper-lined dish, covered with plastic wrap. They can also be frozen for up to 6 months for cooking (throw the frozen berries directly into pancake and muffin batters — don't thaw or they'll collapse).

Blueberries are lovely with lemon, orange, cinnamon, cloves, pecans, walnuts, cream cheese, yoghurt, peach, nectarine, mango and other berries.

Raspberries

With their juicy texture and fine, sweet, musky flavour, these are a late summer fruit to celebrate and savour. There are golden versions of the raspberry, but the best known is the reddish fruit. They are expensive because they are very labour-intensive to pick as they are so fragile.

Use straight away, or refrigerate for no longer than 2 days. Choose dark-coloured berries and try to find organic ones that haven't been sprayed so you won't need to wash them; water diminishes their texture and flavour. Any with green stems attached were picked too early and won't be sweet.

Wonderful on their own, or with whipped cream or crème fraîche, raspberries can be also cooked with a little sugar, in a covered saucepan, just until they release their juices. They also make fabulous jam. They go well with other berries, apple, pear, orange, rhubarb, figs, liqueurs, sauternes, sparkling wine, chocolate (white, milk and dark) and pastry.

Savoury partners include duck, venison, spinach and balsamic vinegar. Raspberry vinegar is a special ingredient used in French savoury cookery.

Strawberries

Their botanical name, *Fragaria*, means 'fragrant' — and that is absolutely what a strawberry should be. Only buy them at the height of summer, as fresh from a good supplier as possible, when they are properly juicy, luscious and rich. When you do find such fruit, they are exquisite on their own, or with no more than a sprinkling of icing (confectioners') sugar and a dollop of whipped cream, lightly sweetened ricotta or mascarpone cheese.

Strawberries won't ripen after picking. They should be plump, fragrant and just a little firm. Their 'shoulders' will be evenly red if ripe; avoid any that are white–green. If you must refrigerate them, place them in a single layer on a tray lined with paper towels, covered tightly with plastic wrap, and bring to room temperature before serving. Don't remove the green leafy hull before rinsing as it stops the flavour dissipating.

Heating destroys the pigment anthocyanin, turning strawberries a dun-coloured mauve. Acid stabilises the pigment, which is why rhubarb, for example, is good to cook with strawberries (and tastes wonderful too).

Besides their obvious uses, strawberries are excellent as a sandwich filling using the freshest white bread, cream cheese and a little chopped mint. They can also be macerated in a little balsamic vinegar, Cointreau or brandy. Strawberries pair well with vanilla, orange, rhubarb, other berries, cream, crème fraîche, ricotta, mascarpone and cream cheese, flower waters (rose and orange), mint, almonds, pistachios and balsamic vinegar.

BROAD (FAVA) BEANS

Fresh broad beans do involve a bit of work to get to the table, but what a treat they are! There is something deeply satisfying about podding and cooking these beans from scratch when they are in season from spring to early summer.

The pods are usually 18–23 cm (7–9 inches) long, each containing 4–8 flattish beans. The pods often look a bit scruffy, but do try to select ones that are smooth and uniformly pale green, with no blackening on the ends. Very bulging pods indicate over-mature beans that will be hard and dry. The best way to check quality is to break a pod open: inside should be tight-skinned, evenly sized pale-green beans nestled in a downy, white lining.

Refrigerate the beans in their pods in a perforated plastic bag for 2–3 days only. They are easy to pod; twist the pod to break it open, peel back the top layer, then remove the beans using your fingers.

The Italians enjoy very young, very fresh broad beans raw, dressed with olive oil, salt and pepper, perhaps with some shaved pecorino or parmesan cheese. Otherwise, once podded, broad beans require cooking. The most common way is to boil them in salted water until fully tender, which can take 10–15 minutes, depending on their size. Once cooked and cool enough to handle, slip off the tough outer skin, then add them at the last minute to stews, soups and stir-fries, or simply reheat in a little butter as a side dish.

Broad beans are excellent in pasta, rice and egg dishes or tossed in a simple salad. They go well with salty hard Italian cheeses, cream, butter, ham, bacon, salmon, prawns (shrimp), crab, baby carrots, peas, artichoke, asparagus, saffron, salad greens, mushrooms and olives. Older, mealier broad beans are excellent puréed, either solo or with potato or celeriac.

FIGS

Steeped in myth and history, these luscious autumn fruits are so rich in minerals they were fed to ancient Greek Olympians for optimum nourishment.

The most common varieties include the **genoa**, with a greenish yellow skin and very sweet amber interior, **black mission**, with a deep purple skin and light pink interior, and **adriatic**, a smaller fig with beautiful green skin and a strawberry-pink interior. All figs share a soft outer skin (although some are thicker and may require peeling), a tender white 'pith' and an interior full of tiny edible seeds, which in a ripe fig are held together in a jelly-like, juicy mass.

Figs are very fragile and perishable, and won't ripen after picking. Ideally they should be eaten directly from the tree, warm from the sun. Ripe figs feel heavy and a little soft, and have slightly bent necks. Avoid any with blemishes, bruises or oozing. Sometimes figs crack open as they mature; such fissures, as long as they are not 'weeping', are quite fine and not a defect. When carting figs home, try to have them packed in tissue paper in a single layer in a box, then refrigerate in a similar fashion for 2 days only. If eating them raw, do bring them to room temperature as chilling dulls their extraordinary flavour.

To prepare, wipe well with a damp cloth and remove any very thick skin using a vegetable peeler. Cut off the stems, avoiding any white liquid that oozes out as it can irritate skin. If cooking figs, keep it simple and brief: grilling, baking and poaching are best. Surplus figs can be made into jam, preserved whole in a rich sugar syrup, or cut up and frozen for 3–4 months (once thawed, they'll really only be good for jams, preserves or baked desserts).

Figs are delicious for antipasti, on pizzas, in salads or as a simple baked accompaniment to roast pork, quail and chicken. They pair beautifully with balsamic vinegar, caramelised onion, fennel seeds, sage, prosciutto, pancetta, cheeses (blue, goat's, ricotta, mascarpone), brandy, honey, almonds, walnuts, pistachios, raspberries, ginger, cinnamon, star anise, saffron and orange.

Peas

The seeds of a legume, peas have historically been more valued in their dried form (as 'green split peas' used in soups) than fresh — Europeans didn't really start eating peas fresh until the late seventeenth century. Frozen peas first came on the scene in the 1920s, and although they are very convenient, they're not as intensely and satisfyingly sweet as freshly podded peas.

Peas really *must* be cooked soon after harvesting: from the moment of picking, their sugars are converted to starch, turning them floury and heavy. Search for them in spring and early summer; greengrocers can't keep them fresh in great quantities. Choose smooth, bright pods that are plump with peas, but not bulging; inside, the peas should be tightly packed and shouldn't rattle. Refrigerate in a perforated plastic bag for 2 days only — or to freeze them, first pod and lightly blanch them, then spread them on a tray in a single layer and freeze until frozen, before transferring to a plastic freezer bag or container for storage. Use within 2 months.

Peas are usually briefly boiled in a minimum of salted water, although the French famously braise them with lettuce. A wonderful side dish or purée, peas are also superb in pasta dishes, soups, risottos and egg-based dishes.

Fresh mint is the classic flavour partner (see page 209), but peas also pair well with tarragon, dill, butter, cream, sour cream, cured meats, chicken, rice, potato, curry spices, prawns (shrimp), lobster and crab.

STONE FRUIT

These soft-fleshed marvels are most luscious at the height of their various seasons, which span the summer months, and for some types, into autumn. Their flesh is fragile so check for bruising or other damage before buying. They are best devoured soon after purchase. If you need to refrigerate them, always bring them back to room temperature to enjoy their exquisite flavours.

Apricots

To taste as it should, an apricot must be picked when fully ripe and nearly bursting with juice. Premature picking, cold storage and transportation ruin its flavour, so do try and source local apricots during their summer peak.

Apricots should smell fragrant, feel heavy and yield a little to the touch, without being soft. Choose those with unblemished, orange-gold skin. Use within a couple of days and store at cool room temperature rather than in the refrigerator, nestled in a single layer in crumpled paper towels.

There is no need to peel apricots. Just before using or eating, simply cut each fruit in half around the stone and the flesh will pull away from the stone.

Over the centuries, apricots have featured prominently in Turkish, Arabic, Persian and Spanish cuisine, particularly in lamb, chicken and rice dishes due to their affinity with sweet spices (cardamom, turmeric, cinnamon, ginger, saffron), and nuts such as almonds, walnuts, hazelnuts and pistachios.

Apricots can be poached, baked, or simmered then puréed, or can even be gently grilled (broiled) or barbecued. Flavour accents include vanilla, amaretto, dessert wines, flower water (orange and rose) and honey. They are divine with yoghurt, crème fraîche, cream and cheeses such as ricotta, mascarpone, camembert and brie. If you have lots of slightly under-ripe fruit on your hands, they'll make excellent jam.

Cherries

So sweet and succulent when they briefly appear at the height of summer, cherries are universally adored. The classic cherry is a rich ruby-red, but some varieties are yellow, with an orange-peachy blush, while others have flesh that is almost white. Cherries probably originated in the Caucasus and Balkan areas; the cherry strudel, black forest cake and chilled cherry soups of countries like Hungary and Germany and the cherry and meat stews of Persia all speak of a long heritage of cooking with cherries.

Highly perishable, cherries are mainly hand harvested as they bruise easily. They should have tight, glossy, bright, unblemished skins. Their skin should be dark for their type, with no splits or oozing. They keep best with their stalks attached; these should be green, not brown. Fully ripe cherries will keep just 2–3 days in a sealed plastic bag in the refrigerator. If you are lucky enough to have a glut of cherries, stone them using a cherry pitter (see page 10) before freezing them, or use them in jams, relishes or chutneys.

Cherries can be softly poached, gently stirred (pitted and raw) into muffin, cake and pudding batters, or cooked and puréed and used in chilled desserts. They pair beautifully with rich meats such as ham, duck, quail and game, and with chocolate, almonds, cinnamon, red wine, cheese (cream, ricotta and mascarpone), star anise, bay leaf, vanilla and brandy.

Peaches and nectarines

These fruits spell 'summer' like few others. A fully ripe peach or nectarine is not a package that travels well, so be suspicious of specimens that have traversed the globe to reach you. Many different types have harvesting periods of a mere few weeks — cultivate your greengrocer so you know what's coming and can plan your cooking (and gorging!) accordingly.

For practical purposes there are really two different kinds of peaches and nectarines, depending on whether the flesh clings to the stone ('cling-stone'), or slips free of it ('free-stone'). If you are cooking them in quantity, free stone varieties will streamline operations, being far easier to prepare.

Choose ones that are heavy, yield slightly to pressure and smell sweet and fragrant. Neither fruit will ripen further once picked. Nectarines don't require peeling, but peaches with very fuzzy skins do — a small sharp paring knife is best for this. Another way to peel either fruit, especially when cooking with them, is to cut a small cross through the skin at the base, then blanch them, a few at a time, in simmering water for 15–20 seconds. Immediately place in a bowl of iced water to cool; the skins should just slip off.

These fruits can be poached whole in a sugar syrup, or chopped and cooked in a minimum of syrup to use in pies, puddings or chilled desserts. They are especially lovely with almonds, pistachio nuts, sparkling wine, rosé, vanilla, raspberry, strawberry, cinnamon, cardamom, yoghurt, cream and crème fraîche. Try them in chutneys, salsas and simple savoury salads.

Plums

In summer and early autumn, this versatile fruit is cheap and plentiful. Its great ability to cross-pollinate has led to a bewildering number of varieties, differing greatly in size, colour and flavour. How to decide which are best?

Buy whichever variety is at its peak, looking for plums that are heavy for their size — especially those covered with a fine, silvery, powder-like bloom, which means they are fresh from the tree. Plums should feel firmish but yield a little when ripe. Avoid any that are soft, very hard, wrinkled, cracked or weeping.

Large, juicy, sweet types are the best for eating out of hand or in fruit salads. They can also be used in baking and dessert recipes — as can more tart sorts, which are good stewed gently with a little sugar, or made into jams, chutneys or preserves (free-stone varieties are the most convenient for these). Plums freeze very well for up to 4 months. Remove skins first by blanching (using the method for peaches and nectarines opposite), and also remove the stones as they can impart a bitter almond flavour.

Plums pair well with sweet spices (cloves, cinnamon, ginger, nutmeg, cardamom, star anise), thyme and rosemary. Made into a relish, pickle, chutney or herb-infused compote, they go well with pork and rabbit terrines and pâtés, ham, roast pork, duck, quail and turkey. Plums make glorious sweet pies, cobblers, crumbles and other hot desserts, where they are delicious with almonds, walnuts, orange, other stone fruit, red wine, brandy, kirsch, balsamic vinegar, cream, sour cream, ricotta and mascarpone cheese.

QUINCE

The quince probably originated in the Middle East, and it is from countries such as Morocco, Turkey and Spain that some of the greatest quince recipes come. A member of the rose family, the quince has a thick, shiny yellow skin that is often covered with a light 'fuzz'. Even though quinces are very hard, even when ripe, they bruise very easily. Their raw flavour is very astringent, but with long, slow cooking (usually after peeling), the pale flesh turns soft and gorgeously, deeply rosy and seductively musky.

Quince season is late autumn through winter. Look for fruit with bright, shiny, yellow skin, without any wrinkling around either end. Quinces keep for up to 2 weeks at cool room temperature, and 3–4 weeks in the refrigerator — seal them in a thick plastic bag or they'll perfume your entire fridge! To prepare, wash well, peel (unless a recipe says not to) using a good, strong peeler, then core and slip immediately into a bowl of acidulated water to stop the flesh browning. Cut out any brown bruised patches as you go.

Quinces are commonly braised, poached or baked in heavy sugar syrup for several hours (to extend the quantity, add some pear or apple). They are fantastic too in jam, jelly, in desserts such as crumbles or charlottes, and as quince paste. In the Middle East and North Africa, they are often used in rich, savoury stews. They pair well with honey, cinnamon, red wine, vanilla, bay leaf, cardamom, ginger, saffron, allspice, cloves and pistachios.

Chicken with apricots and honey

Serves 4

40 g (1½ oz) unsalted butter
1 teaspoon ground cinnamon
1 teaspoon ground ginger
a pinch of cayenne pepper
½ teaspoon freshly ground black pepper
4 x 175 g (6 oz) boneless, skinless chicken
   breasts, trimmed
1 onion, thinly sliced
250 ml (9 fl oz/1 cup) chicken stock
6 coriander (cilantro) sprigs, tied in a
   bunch, plus extra sprigs, to garnish
500 g (1 lb 2 oz) apricots, halved and
   stones removed
2 tablespoons honey
2 tablespoons slivered almonds, toasted
steamed couscous, to serve

Melt the butter in a large frying pan. Add the spices and stir over low heat for 1 minute, or until fragrant. Increase the heat to medium and add the chicken breasts. Cook for 1 minute on each side, taking care not to let the spices burn. Remove the chicken from the pan.

Add the onion to the pan and sauté for 5 minutes, or until softened. Return the chicken to the pan, add the stock and tied coriander sprigs and season with sea salt and freshly ground black pepper. Reduce the heat to low, then cover and simmer for 5 minutes, turning the chicken once.

Transfer the chicken to a serving dish, then cover and leave to rest for 2–3 minutes.

Meanwhile, put the apricots, cut side down, into the pan juices and drizzle with the honey. Cover and simmer for 7–8 minutes, turning the apricots after 5 minutes. Remove the coriander sprigs and discard.

Spoon the apricots and sauce over the chicken. Scatter the almonds over and garnish with a few extra coriander sprigs. Serve with steamed couscous passed separately.

LAMB TAGINE WITH QUINCE
SERVES 4–6

1.5 kg (3 lb 5 oz) boned lamb shoulder, trimmed
2 large handfuls of coriander (cilantro) leaves, chopped, plus extra leaves, to garnish
2 large onions, diced
$\frac{1}{2}$ teaspoon ground ginger
$\frac{1}{2}$ teaspoon cayenne pepper
$\frac{1}{4}$ teaspoon ground saffron threads
1 teasoon ground coriander
1 cinnamon stick
3 tablespoons lemon juice
500 g (1 lb 2 oz) quinces
40 g (1 $\frac{1}{2}$ oz) unsalted butter
100 g (3 $\frac{1}{2}$ oz/ $\frac{1}{2}$ cup) dried apricots
1 tablespoon caster (superfine) sugar
steamed couscous or rice, to serve

Cut the lamb into 3 cm (1 $\frac{1}{4}$ inch) pieces and place in a heavy-based, flameproof casserole dish. Add the coriander, half the onion, the spices and cinnamon stick. Season with sea salt and freshly ground black pepper.

Cover with cold water and bring to the boil over medium heat. Reduce the heat and simmer, partly covered, for 1 $\frac{1}{2}$ hours, or until the lamb is tender.

Meanwhile, put the lemon juice in a large bowl of water. Wash the quinces well. Working with one quince at a time, peel and core the quinces and cut into thick wedges, placing them in the acidulated water to stop them browning as you go.

Melt the butter in a heavy-based frying pan over medium heat. Drain and dry the quince and add to the pan with the remaining onion. Cook for 15 minutes, or until lightly golden, stirring occasionally. After the lamb has been cooking for 1 hour, add the quince mixture, dried apricots and sugar and cook over low heat for 10 minutes for the flavours to merge.

Taste the sauce and adjust the seasoning if necessary. Transfer to a warm serving dish and sprinkle with the extra coriander. Serve with steamed couscous or rice.

LAMB AND ARTICHOKE FRICASSEE
SERVES 8

3 tablespoons lemon juice
6 globe artichokes
2 large tomatoes
3 thyme sprigs
3 parsley sprigs
1 bay leaf
4 tablespoons olive oil
2 kg (4 lb 8 oz) diced lamb
750 g (1 lb 10 oz) brown onions,
   thinly sliced
1 tablespoon plain (all-purpose) flour
2 garlic cloves, crushed
185 ml (6 fl oz/¾ cup) white wine
350 ml (12 fl oz) chicken stock
chopped flat-leaf (Italian) parsley,
   to garnish
lemon wedges, to serve

Add the lemon juice to a large bowl of water. Working with one artichoke at a time, remove the tough outer leaves and trim the stalk to 5 cm (2 inches) long. Peel the stalk using a vegetable peeler. Using kitchen scissors, trim the hard points from the outer leaves, then use a sharp knife to trim the top of the artichoke. Gently open out the leaves in the centre of the artichoke and, using a teaspoon, scrape out the hairy choke. Drop each artichoke into the acidulated water to stop them browning as you go.

Bring a large saucepan of salted water to the boil. Add the artichokes to the boiling water and cook for 5 minutes. Remove using tongs and turn upside down to drain. When cool enough to handle, cut the artichokes into quarters and set aside.

Bring another saucepan of water to the boil. Using a small sharp knife, score a small cross in the base of each tomato. Place the tomatoes in the boiling water for about 20 seconds, remove using a slotted spoon, then plunge into a bowl of iced water. Drain the tomatoes and peel the skins away from the cross. Cut the tomatoes in half, scoop out the seeds with a teaspoon, then roughly chop the flesh.

Form the herbs into a bundle, then tie with kitchen string to make a bouquet garni. Set aside.

Heat half the olive oil in a deep flameproof casserole dish. Add the lamb in batches and fry over medium heat for 5–6 minutes, or until golden, turning occasionally. Remove to a plate.

Heat the remaining oil in the pan, then add the onion and sauté for 8 minutes, or until soft and golden. Add the flour and cook for 1 minute.

Stir in the garlic, tomato, wine and stock, then add the lamb and bouquet garni. Cover and simmer for 1 hour.

Add the artichokes and simmer, uncovered, for a further 15 minutes. Remove the lamb and artichokes using a slotted spoon and place in a serving dish. Keep warm.

Discard the bouquet garni and cook the sauce over high heat until it thickens. Pour the sauce over the lamb and garnish with parsley. Serve with lemon wedges.

GRILLED FIGS IN PROSCIUTTO
MAKES 12

25 g (1 oz) unsalted butter
1 tablespoon orange juice
12 small ripe figs
12 sage leaves
6 slices of prosciutto, trimmed and
   halved lengthways

Melt the butter in a small heavy-based saucepan over low heat and allow to cook for 8–10 minutes, or until the froth subsides and the milk solids appear as brown specks on the bottom of the saucepan.

Strain the butter through a sieve lined with paper towels into a clean bowl, then stir in the orange juice and set aside.

Preheat the oven grill (broiler) to high.

Wash the figs gently and pat dry with paper towels. Starting from the stem end and cutting almost to the base, cut each fig into quarters, taking care not to cut all the way through. Gently open the figs out, like a flower.

Put a sage leaf in the opening of each fig, then wrap a piece of prosciutto around each one, tucking the ends under the base of the fig.

Arrange the figs, cut side up, in a shallow baking dish and brush with the butter mixture. Place the dish under the grill and cook for 1–2 minutes, or until the prosciutto is slightly crisp. Serve warm or at room temperature.

RISI E BISI
SERVES 4–6

1.5 litres (52 fl oz/6 cups) chicken or
  vegetable stock
2 teaspoons olive oil
40 g (1½ oz) unsalted butter
1 small onion, finely chopped
80 g (2¾ oz) pancetta, finely chopped
1 kg (2 lb 4 oz) fresh young peas, shelled
  to give 375 g (13 oz/2½ cups) peas
2 tablespoons chopped flat-leaf (Italian)
  parsley
200 g (7 oz/scant 1 cup) risotto rice
50 g (1¾ oz/½ cup) grated parmesan
  cheese

Pour the stock into a saucepan and bring to the boil.
Reduce the heat, then cover and keep at simmering point.

Heat the olive oil and half the butter in a large heavy-
based saucepan. Add the onion and pancetta and sauté over
medium heat for 5 minutes, or until the onion has softened.

Stir in the peas and parsley and add two ladlefuls of the
stock. Simmer for 6–8 minutes.

Add the rice and the remaining stock and simmer for
12–15 minutes, or until the rice is *al dente*. Stir in the parmesan
and remaining butter, season with sea salt and freshly ground
black pepper and serve.

MINTED PEAS
SERVES 4 AS A SIDE DISH

625 g (1 lb 6 oz/4 cups) peas
4 mint sprigs
30 g (1 oz) unsalted butter
2 tablespoons shredded mint

Put the peas in a saucepan and pour in enough water to
just cover them. Add the mint sprigs. Bring to the boil
and simmer for 5 minutes, or until the peas are just tender.
Drain and discard the mint.

Return the peas to the saucepan, add the butter and
shredded mint and stir over low heat until the butter has
melted. Season with sea salt and freshly ground black pepper
and serve.

CORN AND POLENTA PANCAKES WITH BACON AND MAPLE SYRUP
SERVES 4

90 g (3¼ oz/¾ cup) self-raising flour
110 g (3¾ oz/¾ cup) fine polenta
310 g (11 oz/1½ cups) sweet corn kernels
   (about 3 cobs)
375 ml (13 fl oz/1½ cups) milk
olive oil, for pan-frying
8 slices of rindless bacon
175 g (6 oz/½ cup) maple syrup or
   golden syrup

Preheat the oven to 120°C (235°F/Gas ½).

Sift the flour into a bowl and stir in the polenta. Add the corn and 250 ml (9 fl oz/½ cup) of the milk and stir until just combined. Season with sea salt and freshly ground black pepper, then stir in the remaining milk.

Heat 3 tablespoons olive oil in a large frying pan. Spoon half the batter into the pan to make four 9 cm (3½ inch) pancakes. Cook over medium heat for 2 minutes on each side, or until golden and cooked through. Drain on paper towels and place in the oven to keep warm while cooking the remaining four pancakes, adding more oil if necessary. Transfer to the oven to keep warm.

Add the bacon to the same pan and cook for 5 minutes.

Put two pancakes and two bacon slices on each plate and serve drizzled with maple syrup.

BARBECUED CORN IN THE HUSK
SERVES 8

8 corn cobs, in their husks
125 ml (4 fl oz/½ cup) olive oil,
   plus extra, for brushing
6 garlic cloves, finely chopped
4 tablespoons chopped flat-leaf (Italian)
   parsley
butter, to serve

Peel back the corn husks, leaving them attached to the corn at the end. Pull off the white silks, then wash the corn and pat dry with paper towels.

In a small bowl, mix together the olive oil, garlic, parsley and some sea salt and freshly ground black pepper. Brush each cob with some of the mixture, then pull the husks back up and tie together at the top with string.

Working in batches, steam the corn over a pan of boiling water for 5 minutes, then remove using tongs and pat dry.

Meanwhile, heat a barbecue grill plate or chargrill plate to medium and lightly brush with oil. Barbecue the corn for 20 minutes, turning regularly and occasionally spraying with water to keep the corn moist. Serve hot, with knobs of butter.

Corn and polenta pancakes with bacon and maple syrup

CORN CHOWDER
SERVES 8

90 g (3¼ oz) unsalted butter
2 large onions, finely chopped
1 garlic clove, crushed
2 teaspoons cumin seeds
1 litre (35 fl oz/4 cups) vegetable stock
2 potatoes, peeled and chopped
250 g (9 oz/1 cup) tinned creamed corn
400 g (14 oz/2 cups) corn kernels
　(about 4 cobs)
3 tablespoons chopped parsley
125 g (4½ oz/1 cup) grated cheddar cheese
3 tablespoons pouring (whipping) cream
　(optional)
2 tablespoons snipped chives (optional)

Melt the butter in a large heavy-based saucepan. Add the onion and sauté over medium–high heat for 5 minutes, or until golden.

Add the garlic and cumin seeds and cook for 1 minute, stirring constantly, then pour in the stock and bring to the boil. Add the potato, then reduce the heat and simmer for 10 minutes.

Add the creamed corn, corn kernels and parsley. Bring to the boil, then reduce the heat and simmer for 10 minutes.

Stir in the cheese and season to taste with sea salt and freshly ground black pepper. Stir in the cream, if using, and heat gently until the cheese melts. Serve immediately, sprinkled with snipped chives if desired.

ARTICHOKE, PROSCIUTTO AND ROCKET SALAD
SERVES 4

3 tablespoons lemon juice
4 globe artichokes
2 eggs
3 tablespoons fresh breadcrumbs
3 tablespoons grated parmesan cheese,
   plus extra shaved parmesan, to garnish
olive oil, for pan-frying
8 slices of prosciutto
3 teaspoons white wine vinegar
1 garlic clove, crushed
150 g (5½ oz) rocket (arugula), stalks
   trimmed

Add the lemon juice to a large bowl of water. Working with one artichoke at a time, remove the tough outer leaves and trim the stalk to 5 cm (2 inches) long. Peel the stalk using a vegetable peeler. Using kitchen scissors, trim the hard points from the outer leaves, then use a sharp knife to trim the top of the artichoke. Gently open out the leaves in the centre of the artichoke and, using a teaspoon, scrape out the hairy choke. Drop each artichoke into the acidulated water to stop them browning as you go.

Bring a large saucepan of salted water to the boil. Add the artichokes to the boiling water and cook for 2 minutes. Remove using tongs and turn upside down to drain. When cool enough to handle, cut the artichokes into quarters and set aside.

Whisk the eggs in a bowl. In another bowl, combine the breadcrumbs and grated parmesan, then season with sea salt and freshly ground black pepper. Dip each artichoke quarter into the egg, allowing the excess to drain off, then roll in the breadcrumb mixture to coat.

Fill a heavy-based frying pan with olive oil to a depth of 2 cm (¾ inch) and heat over medium–high heat. Add the artichokes in batches and cook for 2–3 minutes, or until golden, turning once. Remove with a slotted spoon and drain on paper towels.

Heat another tablespoon of olive oil in a non-stick frying pan over medium–high heat. Cook the prosciutto in two batches for 2 minutes each time, or until crisp and golden. Remove from the pan, reserving the oil.

Pour the reserved oil into a small bowl. Add the vinegar and garlic, season lightly and whisk to make a dressing. Put the rocket in a bowl, add half the dressing and toss well.

Divide the rocket, artichoke and prosciutto among serving plates. Drizzle with the remaining dressing, scatter with shaved parmesan and sprinkle with a little sea salt.

ARTICHOKES WITH TARRAGON MAYONNAISE

SERVES 4 AS A STARTER OR AS PART OF AN ANTIPASTI PLATTER

4 globe artichokes
3 tablespoons lemon juice

TARRAGON MAYONNAISE
1 egg yolk
1 tablespoon tarragon vinegar
1/2 teaspoon French mustard
170 ml (5 1/2 fl oz/2/3 cup) olive oil

Working with one artichoke at a time, remove the tough outer leaves and trim the stalk to 5 cm (2 inches) long. Peel the stalk using a vegetable peeler. Using kitchen scissors, trim the hard points from the outer leaves, then use a sharp knife to trim the top of the artichoke. Brush all cut areas of the artichoke with the lemon juice to prevent discolouration. Gently open out the leaves in the centre of the artichoke and, using a teaspoon, scrape out the hairy choke.

Place the artichokes in a steamer and steam over a saucepan of boiling water for 30 minutes, or until tender, adding more boiling water to the pan if necessary. Remove from the heat and set aside to cool to room temperature.

To make the tarragon mayonnaise, put the egg yolk, vinegar and mustard in a bowl and whisk together well. Whisking continuously, add the olive oil, a teaspoon at a time, until the mixture is thick and creamy. As the mayonnaise thickens, add the remaining olive oil in a thin, steady stream, whisking all the while. Season to taste with sea salt and freshly ground black pepper.

Cut each cooled artichoke in half lengthways, then divide among serving plates and serve each with a spoonful of tarragon mayonnaise.

ROMAN-STYLE ARTICHOKES
SERVES 4 AS A STARTER OR AS PART OF AN ANTIPASTI PLATTER

3 tablespoons lemon juice
4 globe artichokes
1 tablespoon toasted fresh breadcrumbs
1 large garlic clove, crushed
3 tablespoons finely chopped parsley
3 tablespoons finely chopped mint
1½ tablespoons olive oil
3 tablespoons dry white wine

Preheat the oven to 190°C (375°F/Gas 5).

Add the lemon juice to a large bowl of water. Working with one artichoke at a time, remove the tough outer leaves and trim the stalk to 5 cm (2 inches) long. Peel the stalk using a vegetable peeler. Using kitchen scissors, trim the hard points from the outer leaves, then use a sharp knife to slice off the top quarter of each artichoke to give a level surface. Gently open out the leaves in the centre of the artichoke and, using a teaspoon, scrape out the hairy choke. Drop each artichoke into the acidulated water to stop them browning as you go.

Put the breadcrumbs, garlic, parsley, mint and olive oil in a bowl, season well with sea salt and freshly ground black pepper and mix together. Fill the centre of each artichoke with the mixture, pressing it in well. Close the leaves as tightly as possible to stop the filling falling out.

Arrange the artichokes, with their stalks facing up, in a deep casserole dish just large enough to fit them all — they should be tightly packed. Sprinkle with sea salt and pour in the wine. Cover with a lid, or a double sheet of kitchen foil secured tightly at the edges. Bake for 1½ hours, or until very tender, checking the artichokes halfway through cooking and adding a little water if necessary so they don't burn.

Serve hot as a first course or side vegetable, or at room temperature as an antipasto.

PLUM SAUCE
MAKES 1 LITRE (35 FL OZ/4 CUPS)

1 large green apple
2 red chillies, seeded and finely chopped
1.25 kg (2 lb 12 oz) ripe, firm, red-fleshed
  plums, halved and stones removed
460 g (1 lb/2½ cups) soft brown sugar
375 ml (13 fl oz/1½ cups) white wine
  vinegar
1 onion, grated
3 tablespoons soy sauce
2 tablespoons fresh ginger, finely chopped
2 garlic cloves, crushed

Peel, core and chop the apple and place in a large heavy-based saucepan with 125 ml (4 fl oz/½ cup) water. Cover and simmer for 10 minutes, or until soft.

Add the remaining ingredients and bring to the boil. Reduce the heat to low and simmer for 45 minutes, stirring often.

Place a coarse sieve over a large bowl, pour in the sauce and push it through the sieve using a wooden spoon. (Alternatively, put the mixture through a food mill.) Discard the solids.

Rinse the pan, then return the sauce to the pan and bring back to the boil. Reduce the heat to medium–high and simmer for 20 minutes, or until the sauce has thickened slightly — it will thicken further on cooling. Pour immediately into hot sterilised jars (see Note below) and seal. Label and date each jar.

Leave for 1 month for the flavours to develop. Store in a cool, dark place for up to 12 months. Once opened, the sauce will keep in the refrigerator for up to 6 weeks.

Plum sauce is delicious with beef or pork spareribs, or Chinese barbecued pork and duck.

NOTE: Jars must always be sterilised before pickles, preserves or jams are put in them for storage, otherwise bacteria will multiply. To sterilise your jars and lids, rinse them with boiling water and place in a warm oven for 20 minutes, or until completely dry. (Jars with rubber seals are safe to warm in the oven and won't melt.) Never dry your jars with a tea towel (dish towel) — even a clean one may have germs on it and contaminate the jars.

ANDALUCIAN ASPARAGUS
SERVES 4 AS A STARTER OR SIDE DISH

500 g (1 lb 2 oz/about 3 bunches)
  asparagus spears
1 thick slice of crusty, country-style bread
3 tablespoons extra virgin olive oil
2–3 garlic cloves
12 blanched almonds
1 teaspoon paprika
1 teaspoon ground cumin
1 tablespoon red wine vinegar or
  sherry vinegar

Trim the woody ends from the asparagus. Remove the crusts from the bread and cut the bread into cubes.

Heat the olive oil in a frying pan. Add the bread, garlic and almonds and sauté over medium heat for 2–3 minutes, or until golden.

Using a slotted spoon, transfer the mixture to a food processor. Add the paprika, cumin, vinegar, 1 tablespoon water and some sea salt and freshly ground black pepper. Process until the mixture is finely chopped.

Return the frying pan to the heat and add the asparagus, with a little extra oil if necessary. Cook over medium heat for 3–5 minutes, or until just cooked, turning often. Transfer to a serving plate.

Add the almond mixture to the pan with 200 ml (7 fl oz) water. Simmer for 2–3 minutes, or until the liquid has thickened slightly. Spoon over the asparagus and serve.

ASPARAGUS PIE
SERVES 6

350 g (12 oz/1¾ cups) plain
(all-purpose) flour
250 g (9 oz) cold unsalted butter,
chopped
170 ml (5½ fl oz/⅔ cup) iced water
1 egg, lightly beaten

FILLING
800 g (1 lb 12 oz) asparagus spears
30 g (1 oz) unsalted butter
½ teaspoon chopped thyme
1 French shallot, chopped
60 g (2¼ oz) sliced ham
80 ml (2½ fl oz/⅓ cup) thick
(double/heavy) cream
2 tablespoons grated parmesan cheese
1 egg
a pinch of ground nutmeg

Sift the flour into a large bowl. Using your fingertips, lightly rub the butter into the flour until the mixture resembles fine breadcrumbs. Make a well in the centre, add almost all the iced water to the well and mix with a flat-bladed knife until the mixture comes together in beads, adding the remaining water if necessary.

Gently gather the dough together, transfer to a lightly floured surface and press into a rectangle measuring about 30 x 12 cm (12 x 5 inches). Fold one end into the centre, then the opposite end over to cover the first. Roll into a rectangle again and repeat the folding three or four times. Wrap in plastic wrap and refrigerate for 45 minutes.

Meanwhile, make the filling. Trim the woody ends from the asparagus. Slice any thick spears in half lengthways. Melt the butter in a large frying pan, then add the asparagus, thyme, shallot and 1 tablespoon water. Cook over medium heat for 3–4 minutes, stirring often, until the asparagus is tender. Season to taste with sea salt and freshly ground black pepper and set aside.

Preheat the oven to 200°C (400°F/Gas 6) and grease a 20 cm (8 inch) fluted, loose-based flan (tart) tin.

Roll the pastry out to a 30 cm (12 inch) circle, then ease it into the prepared tin, leaving the excess hanging over the edge. Place half the asparagus over the pastry, top with the ham, then the remaining asparagus.

In a small bowl, mix together the cream, parmesan, egg and nutmeg. Season well and pour over the asparagus.

Fold the overhanging pastry over the filling, forming loose pleats so the pastry lies flat. Brush the pastry with the beaten egg and bake for 25 minutes, or until golden.

Remove from the oven and allow to cool slightly. Serve warm or at room temperature.

BROAD BEANS WITH HAM
SERVES 4–6

20 g (³/₄ oz) unsalted butter
1 onion, chopped
180 g (6 oz) serrano ham, roughly
  chopped (see Note)
2 garlic cloves, crushed
1.2 kg (2 lb 10 oz) shelled broad (fava)
  beans (see Note)
125 ml (4 fl oz/¹/₂ cup) dry white wine
185 ml (6 fl oz/³/₄ cup) chicken stock
crusty bread, to serve (optional)

Melt the butter in a large saucepan. Add the onion, ham and garlic and sauté over medium heat for 5 minutes, or until the onion has softened.

Add the broad beans and wine and cook over high heat until the liquid has reduced by half. Add the stock, then reduce the heat, cover and simmer for 10 minutes. Remove the lid and simmer for a final 10 minutes.

Serve hot as a vegetable accompaniment, or warm as a snack with crusty bread.

NOTE: Instead of serrano ham, you can use thickly sliced prosciutto — choose one that is pink, soft and sweet, not dry and salty. Use young, tender broad beans for this recipe as older ones will be tougher and require peeling.

BROAD BEAN ROTOLLO WITH WITLOF SALAD
SERVES 4–6

250 g (9 oz/1⅓ cups) shelled broad
   (fava) beans
4 eggs
4 egg yolks
2 teaspoons finely chopped mint
2 teaspoons finely chopped basil
20 g (¾ oz) unsalted butter
80 g (2¾ oz/1 cup) grated pecorino
   cheese

WITLOF SALAD
1 tablespoon chopped basil
4 tablespoons olive oil
2 tablespoons lemon juice
1½ tablespoons pine nuts, toasted
2 baby cos (romaine) lettuces, trimmed,
   washed and dried
2 red witlof (chicory/Belgian endive),
   trimmed, washed and dried

Bring a saucepan of water to the boil. Add the broad beans and a large pinch of sea salt and simmer for 2 minutes. Drain the beans and plunge into iced water. Drain well, then peel.

Preheat the oven to 160°C (315°F/Gas 2–3).

In a bowl, whisk together the eggs, egg yolks, mint and basil. Season with sea salt and freshly ground black pepper.

Melt half the butter in a non-stick frying pan over medium–high heat. Pour in half the egg mixture and cook until the base has set but the top is still a little runny. Scatter half the cheese and half the broad beans over the top.

Slide the omelette from the pan onto a sheet of baking paper. Using the baking paper as a guide, gently roll the omelette into a tight sausage. Roll the baking paper around the omelette and tie both ends with string to stop it unrolling. Place on a baking tray.

Make another omelette with the remaining egg mixture, cheese and broad beans. Roll it up in the same way and place on the baking tray. Transfer to the oven and bake for 8 minutes. Remove from the oven, set aside for 2–3 minutes, then unwrap the omelettes and set aside to cool.

Meanwhile, make the witlof salad. Put the basil, olive oil, lemon juice and 1 tablespoon of the pine nuts in a small food processor or blender and blend until smooth. Season with sea salt and freshly ground black pepper.

Put the cos and witlof leaves in a bowl, drizzle with 2 tablespoons of the dressing and toss to coat. Cut the rotollo into slices and arrange over the salad. Drizzle with the remaining dressing, scatter the remaining pine nuts over the top and serve.

AMARETTI-STUFFED PEACHES
SERVES 6

6 ripe peaches
60 g (2¼ oz) amaretti biscuits, crushed
1 egg yolk
2 tablespoons caster (superfine) sugar,
   plus extra, for sprinkling
3 tablespoons ground almonds
2 teaspoons amaretto
3 tablespoons white wine
20 g (¾ oz) unsalted butter, chopped

Preheat the oven to 180°C (350°F/Gas 4). Lightly grease a 30 x 25 cm (12 x 10 inch) baking dish.

Cut each peach in half and remove the stones; if the peaches are cling-stone, carefully use a paring knife to cut around and remove the stone. Using a paring knife, scoop a little of the flesh out from each peach to create a slight cavity. Chop the scooped-out flesh and place it in a small bowl with the crushed biscuits, egg yolk, sugar, ground almonds and amaretto. Mix together well.

Spoon some of the stuffing mixture into each peach, then place the peaches in the baking dish, cut side up. Sprinkle with the wine and a little extra sugar. Dot with the butter and bake for 20–25 minutes, or until golden. Serve warm.

VARIATION: When in season, you can also use ripe apricots or nectarines for this recipe.

CARAMEL PEACH CAKE
SERVES 10–12

250 g (9 oz) unsalted butter, softened
4 tablespoons soft brown sugar
675 g (1 lb 8 oz/about 5 small) free-stone
  peaches, halved and stones removed
230 g (8 oz/1 cup) caster (superfine) sugar
finely grated zest of 1 lemon
3 eggs
310 g (11 oz/2½ cups) self-raising flour,
  sifted
250 g (9 oz/1 cup) plain yoghurt
1 tablespoon lemon juice

Preheat the oven to 180°C (350°F/Gas 4). Grease a deep, 23 cm (9 inch) round cake tin and line the base with baking paper.

Melt 50 g (1¾ oz) of the butter, then pour it over the base of the prepared cake tin. Sprinkle the brown sugar over. Arrange the peach halves, cut side up, over the base of the tin.

Put the caster sugar, lemon zest and remaining butter in a bowl and beat using electric beaters for 5–6 minutes, or until pale and fluffy. Add the eggs one at a time, beating well after each addition — don't worry if the mixture curdles slightly.

Using a metal spoon, fold in half the flour, then half the yoghurt. Fold in the remaining flour, then the remaining yoghurt. Stir in the lemon juice. Spoon the mixture over the peaches and smooth the surface even.

Bake for 1 hour 25 minutes, or until a skewer inserted into the centre of the cake comes out clean.

Remove from the oven and leave to cool in the tin for 30 minutes, before turning out onto a large serving plate.

Caramel peach cake is best eaten the day it is made.

PEACHES CARDINAL
SERVES 4

4 large, ripe free-stone peaches
300 g (10½ oz/2½ cups) raspberries
3–4 tablespoons icing (confectioners')
   sugar, plus extra, for dusting

If the peaches are very ripe, put them in a bowl and pour boiling water over them. Leave for a minute, then drain and carefully peel. If the peaches are not so ripe, dissolve 2 tablespoons sugar in a saucepan of water, add the peaches, then cover and gently poach for 5–10 minutes, or until tender. Drain and peel.

Let the peaches cool, then cut them in half and remove the stones. Put two halves in each serving glass.

Put the raspberries in a food processor and blend until a fine purée forms, then strain to eliminate the seeds. Alternatively, purée the raspberries by pushing them through a fine sieve.

Sift the icing sugar over the raspberry purée and mix in well. Drizzle the purée over the peaches, then cover and chill thoroughly. Serve dusted with a little icing sugar.

APRICOT COMPOTE
SERVES 4–6

1 orange
1 lemon
1 small vanilla bean, split
110 g (3¾ oz/½ cup) sugar
1 kg (2 lb 4 oz) ripe, firm apricots,
   halved and stones removed
1–2 tablespoons caster (superfine) sugar,
   approximately

Peel two strips of zest, each about 5 cm (2 inches) long, from the orange and lemon. Squeeze all the juice from the orange and 1 tablespoon of juice from the lemon and set aside.

Put the citrus zest in a saucepan with the vanilla bean, sugar and 750 ml (26 fl oz/3 cups) water and bring to the boil. Boil rapidly for 5 minutes.

Put the apricots in a wide saucepan and pour the hot syrup over. Gently bring to the boil, then reduce the heat and simmer for 3–4 minutes, or until the apricots are tender; take care not to overcook them. Transfer the apricots to a bowl using a slotted spoon.

Boil the syrup for a further 10 minutes, or until it reduces and thickens. Remove from the heat, allow to cool for 15 minutes, then stir in the reserved orange and lemon juice. Taste for sweetness and add a little caster sugar if necessary. Strain the syrup over the apricots.

Serve warm or at room temperature.

NECTARINE FEUILLETEES
MAKES 8

2 sheets of frozen butter puff pastry,
    thawed
50 g (1³/₄ oz) unsalted butter, softened
55 g (2 oz/¹/₂ cup) ground almonds
¹/₂ teaspoon natural vanilla extract
5 large free-stone nectarines
3 tablespoons caster (superfine) sugar
100 g (3¹/₂ oz/¹/₃ cup) apricot or peach
    jam, warmed and sieved

Preheat the oven to 200°C (400°F/Gas 6). Line two large baking trays with baking paper.

Cut the pastry sheets into eight 12 cm (4¹/₂ inch) rounds and place on the baking trays.

Put the butter, ground almonds and vanilla in a bowl and stir to form a paste. Evenly spread the paste over the pastry rounds, leaving a 1.5 cm (⁵/₈ inch) border around each.

Cut the nectarines in half, remove the stones, then cut into 5 mm (¹/₄ inch) slices. Arrange the slices over the pastry rounds, overlapping them to cover and leaving a thin border around the edge. Sprinkle the sugar over the nectarines.

Bake for 15 minutes, or until the pastry is puffed and golden and the sugar is bubbling. Remove from the oven and brush the hot nectarines and pastry with the warm jam.

Serve hot or at room temperature.

Nectarine feuilletees are best eaten the day they are made.

CHERRY CLAFOUTIS
SERVES 6–8

30 g (1 oz) unsalted butter, melted
500 g (1 lb 2 oz) cherries, pitted
60 g (2¹/₄ oz/¹/₂ cup) self-raising flour
4 tablespoons sugar
2 eggs, lightly beaten
250 ml (9 fl oz/1 cup) milk
icing (confectioners') sugar, for dusting

Preheat the oven to 180°C (350°F/Gas 4). Brush a 23 cm (9 inch) glass or ceramic shallow pie dish with some of the melted butter.

Spread the cherries in the dish in a single layer.

Sift the flour into a bowl, add the sugar and make a well in the centre. Mix together the eggs, milk and remaining butter, then gradually add to the well, whisking until just combined — do not overbeat or the batter will be tough.

Pour the batter over the cherries and bake for 40 minutes.

Remove from the oven and dust liberally with icing sugar. Serve immediately.

CHERRY PIE
SERVES 6

SWEET SHORTCRUST PASTRY

250 g (9 oz/2 cups) plain (all-purpose) flour

100 g (3½ oz/heaped ¾ cup) icing (confectioners') sugar

125 g (4½ oz) cold unsalted butter, chopped

1 egg yolk, mixed with 1½ tablespoons iced water

1 kg (2 lb 4 oz) cherries, stems removed and pitted

95 g (3¼ oz/½ cup) soft brown sugar

1½ teaspoons ground cinnamon

1–2 drops of natural almond extract

1 teaspoon finely grated lemon zest

1 teaspoon finely grated orange zest

3 tablespoons ground almonds

1 egg, lightly beaten

To make the pastry, sift the flour, icing sugar and a pinch of sea salt into a large bowl. Using your fingertips, lightly rub the butter into the flour until the mixture resembles coarse breadcrumbs. Make a well in the centre. Add the egg yolk mixture to the well and mix using a flat-bladed knife until a rough dough forms. Gently gather the dough together, transfer to a lightly floured surface, then press into a round disc. Cover with plastic wrap and refrigerate for 30 minutes, or until firm.

Roll out two-thirds of the dough between two sheets of baking paper to form a circle large enough to fit a 22 x 20 x 2 cm (8½ x 8 x ¾ inch) pie plate. Remove the top sheet of baking paper and invert the pastry into the pie plate. Cut away the excess pastry using a small sharp knife. Roll out the remaining pastry to make a lid large enough to cover the pie. Cover the pastry and pastry-lined pie plate with plastic wrap and refrigerate for 20 minutes.

Meanwhile, preheat the oven to 190°C (375°F/Gas 5).

Put the cherries, sugar, cinnamon, almond extract and lemon and orange zest in a bowl and mix well.

Sprinkle the ground almonds over the pastry shell. Spoon the cherry filling into the shell, brush the pastry edges with the beaten egg, then cover with the pastry lid. Use a fork to seal the pastry edges together. Cut four slits in the top of the pie to allow steam to escape, then brush the pastry with more beaten egg.

Bake for 1 hour, or until the pastry is golden and the filling is bubbling. Remove from the oven and leave to cool slightly. Serve warm.

Cherry pie is best eaten the day it is made.

CHERRY AND CREAM CHEESE STRUDEL
SERVES 8

250 g (9 oz/1 cup) cream cheese,
  at room temperature
100 ml (3 1/2 fl oz) pouring (whipping)
  cream
1 tablespoon brandy or cherry brandy
1 teaspoon natural vanilla extract
100 g (3 1/2 oz/scant 1/2 cup) caster
  (superfine) sugar
4 tablespoons dry breadcrumbs
4 tablespoons ground almonds
10 sheets of filo pastry
75 g (2 1/2 oz) unsalted butter, melted
425 g (15 oz) cherries, pitted
icing (confectioners') sugar, for dusting

Preheat the oven to 200°C (400°F/Gas 6). Lightly grease a large baking tray.

Put the cream cheese, cream, brandy, vanilla and 3 tablespoons of the sugar in a bowl and beat using electric beaters until smooth.

In another bowl, mix together the breadcrumbs, almonds and remaining sugar.

Lay a sheet of filo pastry on a work surface and cover the remaining sheets with a damp tea towel (dish towel) so they don't dry out. Brush the pastry with some of the melted butter and sprinkle with some of the breadcrumb mixture. Lay another sheet of pastry on top, brush with more butter and sprinkle with more breadcrumbs. Repeat with the remaining filo and breadcrumbs.

Spread the cream cheese mixture evenly over the pastry, leaving a 4 cm (1 1/2 inch) border all around. Arrange the cherries over the cream cheese, then brush some melted butter over the pastry border.

Roll the pastry in from one long side, folding in the ends as you roll. Form into a firm roll and place on the baking tray, seam side down. Brush all over with the remaining butter.

Bake for 10 minutes, then reduce the oven temperature to 180°C (350°F/Gas 4) and bake for a further 30 minutes, or until the pastry is crisp and golden.

Remove from the oven and leave to cool on a wire rack for a few minutes.

To serve, dust liberally with icing sugar and cut into slices using a sharp serrated knife. Serve warm.

Cherry and cream cheese strudel is best eaten the day it is made.

QUINCE PASTE
MAKES ABOUT 1 KG (2 LB 4 OZ/3¼ CUPS)

3 large quinces
800 g (1 lb 12 oz/3²/₃ cups) sugar,
   approximately

Wash the quinces, then place in a saucepan and add enough water to cover. Bring to the boil, then reduce the heat and simmer for 30 minutes, or until tender. Drain and leave until cool enough to handle. Peel and core the quinces, then push them through a sieve, food mill or potato ricer, discarding the solids.

Weigh the quince pulp and place in a heavy-based saucepan. Measure the same weight of sugar as the quince pulp and add to the saucepan. Simmer over low heat, stirring occasionally, for 3½–4½ hours, or until very thick, taking care not to let the mixture burn. Remove from the heat and allow to cool a little.

Line a 28 x 18 cm (11 x 7 inch) rectangular cake tin or dish with plastic wrap, then pour the quince mixture in and leave to cool.

Quince paste can be kept for several months in a tightly sealed container in the refrigerator. Serve with cheese and crackers, or with game such as pheasant.

PLUM COBBLER
SERVES 6

750 g (1 lb 10 oz) plums
4 tablespoons sugar
1 teaspoon natural vanilla extract
whipped cream, to serve (optional)

TOPPING
125 g (4½ oz/1 cup) self-raising flour
60 g (2¼ oz) cold unsalted butter,
  chopped
3 tablespoons soft brown sugar
3 tablespoons milk
1 tablespoon caster (superfine) sugar
icing (confectioners') sugar, for dusting

Preheat the oven to 200°C (400°F/Gas 6).

Cut the plums into quarters and remove the stones. Put the plums, sugar and 2 tablespoons water in a saucepan and bring to the boil, stirring until the sugar has dissolved. Reduce the heat, then cover and simmer for 2 minutes, or until the plums are tender (some varieties will cook more quickly than others). Remove the skins if you wish, then stir in the vanilla. Spoon the mixture into a 750 ml (26 fl oz/3 cup) baking dish.

To make the topping, sift the flour into a large bowl. Using your fingertips, lightly rub in the butter until the mixture resembles fine breadcrumbs. Stir in the brown sugar. Add 2 tablespoons of the milk and mix using a flat-bladed knife until a soft dough forms, adding more milk if necessary.

Turn out onto a lightly floured surface and gather together to form a smooth dough. Roll out until 1 cm (½ inch) thick and cut into rounds using a 4 cm (1½ inch) cutter.

Overlap the rounds around the side of the baking dish, over the filling. Lightly brush with the remaining milk and sprinkle with the caster sugar.

Set the dish on a baking tray and bake for 30 minutes, or until the topping is golden and cooked through.

Serve hot or at room temperature, dusted with icing sugar and with whipped cream, if desired.

Plum cobbler is best eaten the day it is made.

STRAWBERRY CHEESECAKE MUFFINS
MAKES 6

115 g (4 oz/½ cup) caster (superfine)
  sugar
4 tablespoons cream cheese, at room
  temperature
250 g (9 oz/1⅔ cups) strawberries,
  hulled
1 tablespoon strawberry or orange-
  flavoured liqueur
175 g (6 oz/1⅓ cups) plain (all-purpose)
  flour
1 tablespoon baking powder
1 teaspoon finely grated orange zest
½ teaspoon sea salt
20 g (¾ oz) unsalted butter, melted
1 egg
125 ml (4 fl oz/½ cup) milk
icing (confectioners') sugar, for dusting

Preheat the oven to 180°C (350°F/Gas 4). Lightly grease a six-hole non-stick muffin tin.

Put half the sugar in a bowl, add the cream cheese and mix together well. Set aside.

Set aside six small strawberries. Place the rest in a blender or food processor with the liqueur and remaining sugar. Blend to a smooth sauce, then strain through a fine sieve to remove the strawberry seeds. Set the strawberry sauce aside for serving with the muffins.

Sift the flour and baking powder into a large bowl, then stir in the orange zest and sea salt. In a separate bowl, beat the butter, egg and milk together, then add to the dry ingredients and mix until just combined — do not overmix or the muffins will be tough.

Spoon half the batter into the muffin holes, then add a reserved strawberry and a teaspoon of the cream cheese mixture to each one. Top with the remaining batter and bake for 15 minutes, or until cooked through and golden. To test if the muffins are cooked, insert a cake tester into them, avoiding the cream cheese mixture — the tester should withdraw clean.

Remove the muffins from the tins and allow to cool slightly. Serve dusted with icing sugar and drizzled with the strawberry sauce.

Strawberry cheesecake muffins are best eaten the day they are made.

Strawberry and mascarpone tart
Serves 6

185 g (6½ oz/1½ cups) plain
  (all-purpose) flour
125 g (4½ oz) cold unsalted butter,
  chopped
4 tablespoons iced water
500 g (1 lb 2 oz) strawberries, hulled
  and halved
2 teaspoons natural vanilla extract
50 ml (1½ fl oz) Drambuie, Cointreau
  or Grand Marnier
4 tablespoons soft brown sugar
250 g (9 oz) mascarpone cheese
300 ml (10 fl oz) thick (double/heavy)
  cream
2 teaspoons finely grated orange zest

Sift the flour and a pinch of sea salt into a large bowl. Using your fingertips, lightly rub the butter into the flour until the mixture resembles fine breadcrumbs. Make a well in the centre, then add almost all the iced water to the well. Mix using a flat-bladed knife until a rough dough forms, adding the remaining water if necessary. Gently gather the dough together, transfer to a lightly floured surface, then press into a round disc. Cover with plastic wrap and refrigerate for 30 minutes, or until firm.

Roll the dough out between two sheets of baking paper until large enough to line a lightly greased 23 cm (9 inch) loose-based tart tin. Trim the excess pastry using a small sharp knife, then refrigerate the pastry-lined tin for 15 minutes.

Meanwhile, preheat the oven to 200°C (400°F/Gas 6) and place a baking tray in the oven to heat.

Line the pastry shell with baking paper and half-fill with baking beads, dried beans or rice. Place on the heated baking tray and bake for 15 minutes, then remove the paper and baking beads and bake for a further 10–15 minutes, or until the pastry is dry and golden. Remove from the oven and allow to cool completely.

In a bowl, mix together the strawberries, vanilla, liqueur and 1 tablespoon of the sugar. In another bowl, mix together the mascarpone, cream, orange zest and remaining sugar. Cover both bowls and refrigerate for 30 minutes, gently tossing the strawberries once or twice.

Whip half the mascarpone mixture until firm, then evenly spoon it into the pastry shell.

Drain the strawberries, reserving the liquid. Pile the strawberries onto the tart.

Slice the tart and serve with a drizzle of the reserved strawberry liquid and the remaining mascarpone cream.

Strawberry and mascarpone tart is best eaten the day it is made.

STRAWBERRIES WITH BALSAMIC VINEGAR
SERVES 4–6

750 g (1 lb 10 oz) ripe small strawberries
3 tablespoons caster (superfine) sugar
2 tablespoons good-quality balsamic
　vinegar
mascarpone cheese, to serve

Wipe the strawberries with a clean damp cloth and hull them. Cut any large strawberries in half lengthways.

Place the strawberries in a glass bowl, sprinkle the sugar over the top and toss gently to coat. Cover with plastic wrap and leave for 30 minutes to macerate. Sprinkle the vinegar over the strawberries, toss gently, then cover and refrigerate for 30 minutes.

Divide the strawberries among serving glasses, drizzle with the syrup and serve with a dollop of mascarpone cheese.

WATERMELON GRANITA
SERVES 4

450 g (1 lb) watermelon, rind and seeds
　removed
1 tablespoon caster (superfine) sugar
1/2 teaspoon lemon juice

Purée the watermelon flesh in a blender or food processor, or chop it finely and push it through a sieve.

Put the sugar, lemon juice and 75 ml (2 1/2 fl oz) water in a small saucepan, then stir over low–medium heat for 4 minutes, or until the sugar has dissolved. Remove from the heat, add the watermelon purée and stir until well combined.

Pour the mixture into a deep ceramic or glass dish. Cover with plastic wrap and freeze, stirring the mixture every 30 minutes with a fork to break up the ice crystals and refine the texture.

Just before serving, stir the granita with a fork to break up the ice crystals. Granita will keep, frozen, for up to 5 days.

Strawberries with balsamic vinegar

FIGS IN HONEY SYRUP
SERVES 4–6

12 whole fresh figs
100 g (3½ oz/⅔ cup) blanched whole
  almonds, lightly toasted
110 g (3¾ oz/½ cup) sugar
4 tablespoons honey
2 tablespoons lemon juice
a 6 cm (2½ inch) strip of lemon zest
1 cinnamon stick
250 g (9 oz/1 cup) Greek-style yoghurt

Cut the stems from the figs and make a small crossways incision 5 mm (¼ inch) deep on top of each. Push a toasted almond into each fig, through the incision.

Pour 750 ml (26 fl oz/3 cups) water into a large saucepan, add the sugar and stir over medium heat until the sugar has dissolved. Increase the heat and bring to the boil.

Stir in the honey, lemon juice and lemon zest and add the cinnamon stick. Reduce the heat to medium, add the figs and gently simmer for 10 minutes. Remove the figs with a slotted spoon and place in a serving dish.

Boil the liquid over high heat for 15–20 minutes, or until thick and syrupy. Remove the cinnamon and lemon zest. Allow the syrup to cool slightly, then pour over the figs.

Roughly chop the remaining almonds and sprinkle over the figs. Serve warm or cold, with the yoghurt.

FIG AND RASPBERRY SHORTCAKE
SERVES 6

185 g (6½ oz) unsalted butter, softened
170 g (6 oz/¾ cup) caster (superfine)
  sugar
1 egg
1 egg yolk
335 g (11¾ oz/2⅔ cups) plain
  (all-purpose) flour
1 teaspoon baking powder
4 figs, trimmed and quartered
200 g (7 oz/1⅔ cups) raspberries
finely grated zest of 1 orange
2 tablespoons sugar
whipped cream or mascarpone cheese,
  to serve

Put the butter and caster sugar in a bowl and beat using electric beaters until pale and fluffy. Add the egg and egg yolk and beat well.

Sift the flour, baking powder and a pinch of sea salt into the bowl, then gently fold into the butter mixture. Cover with plastic wrap and refrigerate for 15 minutes, or until firm enough to roll out.

Meanwhile, preheat the oven to 180°C (350°F/Gas 4). Lightly grease a 23 cm (9 inch) spring-form cake tin.

Divide the dough in half. On a lightly floured surface, roll out one portion until it is large enough to fit the base of the cake tin. Fit it into the tin and arrange the figs and raspberries over the top. Sprinkle with the orange zest.

Roll out the remaining dough to the same size and fit it over the filling. Lightly brush the top with water and sprinkle with the sugar.

Bake for 35 minutes, or until the pastry is golden. Remove from the oven and allow to cool a little.

Serve warm or at room temperature, with whipped cream or mascarpone cheese.

Fig and raspberry shortcake is best eaten the day it is made.

AMARETTI, PEAR AND RASPBERRY TRIFLE
SERVES 4

170 g (6 oz/^{3}/$_4$ cup) caster (superfine)
  sugar
2 ripe, firm pears, such as packham or
  williams, peeled, halved and cored
1 tablespoon marsala
2 teaspoons instant coffee granules
16 amaretti biscuits, roughly broken
2 tablespoons orange juice
200 g (7 oz/1^2/$_3$ cups) raspberries
vanilla ice cream, to serve

CUSTARD
420 ml (14^1/$_2$ fl oz/1^2/$_3$ cups) skim milk
2 tablespoons caster (superfine) sugar
1 teaspoon natural vanilla extract
2^1/$_2$ tablespoons custard powder or instant
  vanilla pudding mix

Put the sugar in a saucepan with 375 ml (13 fl oz/1^1/$_2$ cups) water. Slowly bring the mixture to a simmer, stirring occasionally to dissolve the sugar. Add the pears and cook over low–medium heat for 10 minutes, or until the pears are tender. Drain the pears well, discarding the syrup, and set aside.

To make the custard, put the milk, sugar and vanilla in a heavy-based saucepan and bring nearly to a simmer over low heat, stirring occasionally. Mix the custard powder with 2 tablespoons water to form a smooth paste, then whisk into the milk mixture until the custard boils and thickens. Remove from the heat and cover the surface directly with plastic wrap to prevent a skin forming. Set aside to cool.

Put the marsala and coffee granules in a small bowl and stir to dissolve the coffee.

Place the biscuits and orange juice in a large bowl and stir to coat the biscuits. Layer half the biscuits in four serving glasses and drizzle with the marsala mixture. Top with a third of the raspberries.

Roughly chop the pears and divide half among the serving glasses. Pour in half the custard.

Repeat the layering, finishing with the raspberries.

Chill for 10 minutes, or serve immediately with a scoop of vanilla ice cream.

BLUEBERRY CHEESECAKE
SERVES 8–10

125 g (4¹/₂ oz) unsalted butter
100 g (3¹/₂ oz/1 cup) rolled (porridge)
  oats
100 g (3¹/₂ oz) wheatmeal biscuits, finely
  crushed
2 tablespoons soft brown sugar
250 g (9 oz/1²/₃ cups) fresh blueberries
240 g (8¹/₂ oz/³/₄ cup) blackberry jam
3 tablespoons cherry brandy

FILLING
375 g (13 oz/1¹/₂ cups) cream cheese,
  at room temperature
100 g (3¹/₂ oz/heaped ¹/₃ cup) ricotta
  cheese
4 tablespoons caster (superfine) sugar
125 g (4¹/₂ oz/¹/₂ cup) sour cream
2 eggs
1 tablespoon finely grated orange zest
1 tablespoon plain (all-purpose) flour

Grease a deep, 20 cm (8 inch) round spring-form cake tin and line the base with baking paper.

Melt the butter in a saucepan, add the oats and biscuit crumbs and mix well. Stir in the sugar.

Press half the biscuit mixture firmly over the base of the cake tin, then gradually press the remainder around the side of the tin, using a glass to firm it into place, and pressing it about three-quarters of the way up to the edge. Refrigerate for 10–15 minutes, or until firm.

Meanwhile, preheat the oven to 180°C (350°F/Gas 4).

To make the filling, put the cream cheese, ricotta, sugar and sour cream in a bowl and beat using electric beaters until smooth. Beat in the eggs, orange zest and flour.

Set the cake tin on a baking tray, then pour the filling into the crust. Transfer the cake tin on the baking tray to the oven and bake for 40–45 minutes, or until the filling is just set. Remove from the oven and leave in the tin to cool.

Arrange the blueberries over the cooled cheesecake.

Put the jam and cherry brandy in a small saucepan and gently heat until the jam has melted. Simmer for 1–2 minutes, then remove from the heat and push the mixture through a sieve into a bowl. Allow to cool slightly, then carefully brush the mixture over the blueberries.

Refrigerate the cheesecake for several hours or overnight before serving.

Blueberry cheesecake will keep refrigerated for 2 days.

Berries in champagne jelly
Serves 8

1 litre (35 fl oz/4 cups) Champagne or
   sparkling white wine
1½ tablespoons powdered gelatine
220 g (7¾ oz/1 cup) sugar
4 lemon zest strips
4 orange zest strips
250 g (9 oz/1⅔ cups) small strawberries,
   hulled and halved
250 g (9 oz/1⅔ cups) blueberries

Pour 500 ml (17 fl oz/2 cups) of the Champagne into a bowl and let the bubbles subside. Sprinkle the gelatine evenly over the Champagne, then leave to stand for 5 minutes, or until the gelatine has softened.

Pour the remaining Champagne into a large saucepan. Add the sugar, lemon zest and orange zest. Heat gently, stirring constantly for 3 minutes, or until the sugar has dissolved.

Remove the pan from the heat, add the gelatine mixture and stir until dissolved. Set aside to cool, then remove the lemon and orange zest.

Divide the berries among eight small wine glasses or martini glasses and carefully pour the jelly over the top. Refrigerate for 6 hours or overnight, or until the jelly has set.

Remove from the refrigerator 15 minutes before serving.

Raspberry jam
Fills six 250 ml (9 fl oz/1 cup) jars

1.5 kg (3 lb 5 oz) raspberries
4 tablespoons lemon juice
1.5 kg (3 lb 5 oz) sugar

Put the raspberries and lemon juice in a large heavy-based saucepan. Gently cook over low heat for 10 minutes, or until the raspberries have softened, stirring occasionally.

Meanwhile, warm the sugar by spreading it in a large baking dish and heating it in a 120°C (250°F/Gas ½) oven for 10 minutes, stirring occasionally.

Add the sugar to the pan and stir, without boiling, for 5 minutes, or until the sugar has completely dissolved.

Put two small plates in the freezer. Bring the jam to the boil and boil for 20 minutes, then start testing for setting point by placing a little of the hot jam on a chilled plate. When setting point is reached, a skin will form on the surface and the jam will wrinkle when you push it with your finger. If the jam doesn't set, keep cooking and testing until it does — this may take up to 40 minutes.

Allow to cool for 5 minutes, then skim off any impurities that have risen to the surface. Pour into hot sterilised jars (see Note below) and seal. Allow to cool completely, then label and date each jar.

Store in a cool, dark place for 10–12 months. Once opened, raspberry jam will keep in the refrigerator for up to 4 weeks.

NOTE: Jars must always be sterilised before pickles, preserves or jams are put in them for storage, otherwise bacteria will multiply. To sterilise your jars and lids, rinse them with boiling water and place in a warm oven for 20 minutes, or until completely dry. (Jars with rubber seals are safe to warm in the oven and won't melt.) Never dry your jars with a tea towel (dish towel) — even a clean one may have germs on it and contaminate the jars.

RHUBARB AND BLACKBERRY CRUMBLE
SERVES 4

850 g (1 lb 14 oz) rhubarb, cut into
  2.5 cm (1 inch) lengths
150 g (5½ oz/1¼ cups) blackberries
1 teaspoon grated orange zest
230 g (8 oz/1 cup) caster (superfine) sugar,
  plus extra, if necessary
125 g (4½ oz/1 cup) plain (all-purpose)
  flour
115 g (4 oz/1 heaped cup) ground
  almonds
½ teaspoon ground ginger
150 g (5½ oz) cold unsalted butter,
  chopped
cream or ice cream, to serve

Preheat the oven to 180°C (350°F/Gas 4). Grease a deep 1.5 litre (52 fl oz) baking dish.

Put the rhubarb in a large saucepan with 2½ tablespoons water. Cover and cook over medium heat for 15 minutes, or until the rhubarb has softened, shaking the pan occasionally. Remove from the heat and stir in the blackberries, orange zest and 4 tablespoons of the sugar. Taste and add a little more sugar if necessary. Spoon the mixture into the baking dish.

Sift the flour into a bowl and mix in the ground almonds, ginger and remaining sugar. Using your fingertips, rub the butter into the flour mixture until it resembles coarse breadcrumbs. Sprinkle the mixture over the fruit, pressing it down lightly — don't press too firmly or the crumble will be flat and dense.

Place the baking dish on a baking tray, then transfer to the oven and bake for 25–30 minutes, or until the topping is golden and the fruit is bubbling underneath.

Remove from the oven and leave to cool for 5 minutes. Serve hot or warm, with cream or ice cream.

Rhubarb and blackberry crumble is best eaten the day it is made.

VARIATION: You can substitute raspberries, loganberries or blueberries for the blackberries.

BLACKBERRY PIE
SERVES 6

250 g (9 oz/2 cups) plain (all-purpose) flour
100 g (3½ oz/heaped ¾ cup) icing (confectioners') sugar
125 g (4½ oz) cold unsalted butter, chopped
1 egg yolk, mixed with 1½ tablespoons iced water
500 g (1 lb 2 oz) blackberries
145 g (5 oz/⅔ cup) caster (superfine) sugar, plus extra, for sprinkling
2 tablespoons cornflour (cornstarch)
milk, for brushing
1 egg, lightly beaten

Sift the flour, icing sugar and a pinch of sea salt into a large bowl. Using your fingertips, lightly rub the butter into the flour mixture until the mixture resembles coarse breadcrumbs. Make a well in the centre. Add the egg yolk mixture to the well and mix using a flat-bladed knife until a rough dough forms. Gently gather the dough together, transfer to a lightly floured surface, then press into a round disc. Cover with plastic wrap and refrigerate for 30 minutes, or until firm.

Meanwhile, preheat the oven to 200°C (400°F/Gas 6) and grease a 20 cm (8 inch) pie dish.

Roll out two-thirds of the pastry between two sheets of baking paper until large enough to line the dish, pressing it into place and trimming the edges with a small sharp knife.

Put the blackberries, caster sugar and cornflour in a bowl and gently toss until well coated. Spoon into the pastry shell.

Roll out the remaining pastry between two sheets of baking paper until large enough to cover the pie. Moisten the rim of the pie base by brushing it with milk, then press the pastry lid firmly into place. Trim and crimp the edges. Brush with the beaten egg and sprinkle with a little extra caster sugar. Pierce the top of the pie with a knife.

Bake on the bottom shelf of the oven for 10 minutes. Reduce the oven temperature to 180°C (350°F/Gas 4) and move the pie to the centre shelf. Bake for a further 30 minutes, or until golden on top. Remove from the oven and allow to cool before serving.

Blackberry pie is best eaten the day it is made.

Sweet grape flatbread

Serves 6–8

100 g (3½ oz/¾ cup) raisins
90 ml (3 fl oz) sweet marsala
150 ml (5 fl oz) milk
115 g (4 oz/½ cup) caster (superfine)
    sugar
2 teaspoons active dried yeast
300 g (10½ oz/scant 2½ cups) plain
    (all-purpose) flour, plus extra, for
    dusting
400 g (14 oz/2¼ cups) black seedless
    grapes

Put the raisins in a bowl and pour the marsala over.
Set aside.

Warm the milk and pour into a small bowl. Stir in
1 teaspoon of the sugar, sprinkle the yeast over and set aside
in a draught-free place for 10 minutes, or until foamy.

Put the flour, 4 tablespoons of the sugar and a pinch
of sea salt in a bowl and mix together. Add the yeast mixture
and mix using a wooden spoon until a rough dough forms.

Turn out onto a lightly floured surface and knead for
6–8 minutes, or until smooth and elastic. Add a little more
flour or a few drops of warm water if necessary to give a soft,
but not sticky, dough.

Place the dough in a large oiled bowl, turning to coat
in the oil. Cover with plastic wrap and leave to rise in a
draught-free place for 1 hour, or until doubled in size.

Drain the raisins and squeeze them dry. Lightly dust a
baking tray with flour. Gently deflate the dough using a lightly
floured fist, then divide in half. Shape each half into a flattened
round about 20 cm (8 inches) in diameter and place one round
on the baking tray. Scatter half the grapes and half the raisins
over the dough, then cover with the second round of dough.
Scatter the remaining grapes and raisins over the top. Cover
loosely with a tea towel (dish towel) and leave in a draught-
free place for 1 hour, or until doubled in size.

Meanwhile, preheat the oven to 180°C (350°F/Gas 4).

Sprinkle the dough with the remaining sugar and bake
for 40–50 minutes, or until golden.

Serve warm or at room temperature, cut into thick slices.

index

Published in 2008 by Murdoch Books Pty Limited
www.murdochbooks.com.au

Murdoch Books Australia
Pier 8/9
23 Hickson Road
Millers Point NSW 2000
Phone: +61 (0) 2 8220 2000
Fax: +61 (0) 2 8220 2558

Murdoch Books UK Limited
Erico House
6th Floor
93–99 Upper Richmond Road
Putney, London SW15 2TG
Phone: +44 (0) 20 8785 5995
Fax: +44 (0) 20 8785 5985

Chief Executive: Juliet Rogers
Publishing Director: Kay Scarlett

Design Manager: Vivien Valk
Project Manager: Janine Flew
Editor: Katri Hilden
Design concept: Sarah Odgers
Design: Alex Frampton
Production: Kita George
Photographer: George Seper
Stylist: Marie-Helénè Clauzon
Food preparation: Joanne Glynn

National Library of Australia Cataloguing-in-Publication Data
Author: Kitchen, Leanne.
Title: The greengrocer / Leanne Kitchen; editor, Janine Flew.
ISBN: 9781741962000 (hbk.)
Series: Providore series
Notes: Includes index.
Subjects: Fruit. Vegetables. Herbs.
Other Authors/Contributors: Flew, Janine.
Dewey Number: 634

Colour separation by Splitting Image. Printed by 1010 Printing International Limited.
PRINTED IN CHINA

CONVERSION GUIDE: You may find cooking times vary depending on the oven you
are using. For fan-forced ovens, as a general rule, set the oven temperature to 20°C (35°F)
lower than indicated in the recipe.